R. S. Jackson, PhD

Conserve Water, Drink Wine: Recollections of a Vinous Voyage of Discovery

Pre-publication REVIEWS, COMMENTARIES, EVALUATIONS . . .

"This is a delightful, informative, joyful book! Knowledgeable oenophiles as well as neophytes will find this to be a highly rewarding, entertaining, and instructive 'voyage.' Before you realize it, you have learned things about wine the easy way–quite sophisticated things, discussed in a conversational manner by people who know what they are talking about. How many times have you hestitated to ask a question about wine history, winemaking, wine chemistry, or wine and food combinations because you were not sure that the question wasn't 'silly?' Well, in this book the author lets his friends ask the questions, and experts answer them as though they were having a conversation around your dinner table. From botrytis to phylloxera, from carbonic maceration to malolactic fermentation, from the real meaning of 'moderate' wine consumption to the health benefits thereof, you learn something new every time.

Join Ron Jackson and trudge the vineyards, explore the wineries, assess the wines, eavesdrop on the kind of dinner conversation only a well-set table and a well-chosen wine accompaniment can generate, and you will realize that the enjoyment of this elixir makes you a part of a vinous voyage of discovery that began six thousand years ago and still yields its happy surprises."

H. J. de Blij, PhD
University Scientist,
University of South Florida,
Boca Grande, FL

More pre-publication
REVIEWS, COMMENTARIES, EVALUATIONS . . .

"**R**on Jackson has written a superbly instructive narrative on winemaking and grape growing. His book covers all the bases. With so much detail, it could be used as a textbook for a wine class. I was particularly impressed with the up-to-date information in the wine and health chapter."

Andrew L. Waterhouse, PhD
Assistant Professor,
Department of Viticulture
and Enology,
University of California, Davis

"**T**his ambitious book seeks to distill a wide range of scientific information on viticulture and winemaking into a readable form for wine enthusiasts without a scientific background. Its title, *Conserve Water, Drink Wine*, is taken from the bumper sticker on the author's station wagon, and reflects his slightly anarchic approach. It is written in the form of a series of lively discussions between a group of wine enthusiasts with whom the author meets up on a sabbatical in Ithaca, New York. Although these discussions are surely too contrived to have been real, they introduce the reader in a very easy way to many of the complexities of grape growing and winemaking, while also including a wealth of fascinating anecdotal material. The friends meet up at home, in vineyards, wineries, and research stations; have visitors from abroad come to talk to them about their particular expertises; and conclude with a meal at which they chat about food and wine combinations.

Also the author of *Wine Science: Principles and Applications*, Ron Jackson knows the relevant scientific literature well. He has managed to include within this new book a tremendous amount of material, including the findings of much recent research, and has presented it all in a novel and at times highly amusing manner. Read a chapter at a time, with a glass of suitable wine in hand, it provides a very easy way of acquiring a considerable amount of useful knowledge about wine. Although the location of the author's journey is the Finger Lakes Region of New York, the group's discussions take them far away from there to wines made in many other areas of the world, and particularly the traditional European heartland of winemaking.

Overall, Ron Jackson has produced an unusual, informative, and highly readable book that provides an easily accessible source of knowledge about many aspects of grape growing and winemaking."

Tim Unwin, PhD
Reader in Geography,
Royal Holloway,
University of London

More pre-publication
REVIEWS, COMMENTARIES, EVALUATIONS . . .

"**C**onserve Water, Drink Wine* is a consumer's textbook for wine enjoyment. It presents all the wine science any neophyte might want to know, yet does it in a way that allows the reader to pick and choose his or her own balance between objective wine chemistry and subjective wine appreciation. The reader is led on a first-time adventure through New York State's famous Finger Lakes wine country. The book reads as a diary of wine country experiences in which knowledgeable friends and acquaintances explain many technical details about wine growing, winemaking, and wine appreciation.

The education process begins with a wine tasting and discussion evening in which wine history merges with sensory evaluation. Visiting a local vineyard then allows the reader to experience home winemaking firsthand while learning basic fundamentals of yeast fermentation and wine chemistry. Then they visit selected commercial wineries, getting a little deeper into the science of wine and learning that 'everybody does it a little differently in practice.'

Finally, there are discussions with experts who round out the whole experience by filling in the inevitable holes that had to be left earlier. One of the book's features is that the author studies and teaches wine without making it something to worship pompously or to gaze at with awe or mysticism.

This book will serve anyone well who wants to develop a knowledge of wine appreciation–from growing through winemaking, aging, and consumption. Best of all, it encourages the reader to continue learning more about the excitement of wine after the final page is turned."

Richard G. Peterson, PhD
*Chairman, Folie à Deux Winery,
St. Helena, Napa Valley, California*

"**T**his book presents many aspects of wine production and enjoyment, but avoids the dry technical presentation one might expect in a textbook. Written around a sabbatical visit to Cornell University, the book details how a group of acquaintances become friends as they learn about wine together.

Jackson records all discussions of the group in a conversational format so that the reader can feel like part of the group in attendance. For example, a knowledgeable local wineshop owner talks about wine selection in a fashion designed to demystify the process. Another member of the group is of Italian origin, and he teaches them the joys of grape harvesting and elements of winemaking in his well-equipped garage!

A visitor to the group is a medical allergist who discusses wine and health, again in terms that are easy to understand. The Finger Lakes Region of Upstate New York is an ideal setting for the book, allowing for visits to the Gold Seal Winery and the experimental vineyards of Geneva Experiment Station, and for a meal prepared by students of Cornell University's famous Hotel School. The text is broken up by diagrams and cartoons, the latter designed to present a less serious side of wine.

This book is ideal for readers who want to learn more about wine and are not afraid of some basic technical information that will help their wine enjoyment. The author credits his wife Suzanne with suggesting a book of this type; many readers will be glad for her encouragement."

Richard Smart, Bsc,Msc,PhD
International Viticultural Consultant, Smart Viticultural Services

Food Products Press
An Imprint of The Haworth Press, Inc.

Conserve Water, Drink Wine
Recollections of a Vinous Voyage of Discovery

Conserve Water, Drink Wine
Recollections of a Vinous Voyage of Discovery

R. S. Jackson, PhD

Graphics by
Herman Casteleyn

Food Products Press
An Imprint of The Haworth Press, Inc.
New York • London

Published by

Food Products Press, an imprint of The Haworth Press, Inc., 10 Alice Street, Binghamton, NY 13904-1580

Cover design by Donna M. Brooks.

Library of Congress Cataloging-in-Publication Data

Jackson, Ron S.
    Conserve water, drink wine : recollections of a vinous voyage of discovery / by R. S. Jackson; graphics by Herman Casteleyn.
       p.   cm.
    Includes bibliographical references and index.
    ISBN 1-56022-864-4 (alk. paper)
    1. Wine and wine making. I. Title.
TP548.J147 1997
641.2′2–dc20   96-9551
                                                         CIP

This book is dedicated to my wife, Suzanne Ouellet, who has been the inspiration and most constructive critic of this work; to Marvin Myers, whose fine wit has probed the humorous world of wine; and to those ever marvelous microbes that mysteriously metamorphose grapes into wine.

ABOUT THE AUTHOR

Ron S. Jackson, PhD, is Professor and Chairman of the Botany Department at Brandon University in Brandon, Manitoba. He teaches a wide variety of Applied Microbiology and Botany courses and one of the few courses in Wine Technology offered in Canada. Dr. Jackson is the author of *Wine Science: Principles and Applications,* which received honorable mention from the Association of American Publishers. An apprenticeship in Vineland, Ontario, and a sabbatical at Cornell University redirected his interest in *Botrytis*-induced plant diseases to a thirst for wine knowledge. He serves as a consultant for the Manitoba Liquor Control Commission.

CONTENTS

Preface   **xi**

Acknowledgments   **xiii**

Chapter 1. Arrival in Ithaca   **1**

Chapter 2. The CWDW Society   **11**

Origin of Winemaking   13
Domestication of the Grapevine   16
Wine Sampling and Terminology   18

Chapter 3. Visit to Venture Vineyards and Home Winemaking   **27**

Influence of Slope   28
Influence of Soil Texture and Pruning   29
Influence of Adjacent Water Bodies   32
Timing of Harvest   33
Harvest Method   35
Carbonic Maceration Winemaking   36
Yeast Inoculation   40
Traditional White Winemaking   40
Traditional Red Winemaking   45
Decanting   47

Chapter 4. Visit to Heron Hill Winery   **55**

Stemmer/Crusher   57
Presses   60
Use of Sulfur Dioxide   60
Fermentors   62
Chaptalization   66
Yeasts and Fermentation   67
Malolactic Fermentation   68
Maturation in Oak   70

Chapter 5. Views on Wine Selection   **75**

Types of Consumers   79
Vintage   82
Appellation Control Laws   83
Varietal Origin   85
Geographic Origin   86
Aging Potential Indicators   88
Bag-in-Box   90
Label Removal   92

Chapter 6. Bacchanalian Pleasures and Faults   **93**

Wine Glasses   93
Corkscrews   96
Wine Appearance   98
Wine Fragrance   100
Wine Taste   105
Wine Tears   108
Wine Flavor   108
Wine Faults   110

Chapter 7. Thoughts on Wine Quality, Aging, and Fraud Detection   **115**

Concepts of Wine Quality   116
Breathing   120
Wine Aging   121
Wine Fraud   128

Chapter 8. Visit from Dr. Nicholson: Wine and Health   **133**

Wine-Induced Headaches   134
Effect on Food Digestion   137
Wine and Arteriosclerosis   137
Wine's Antioxidant Effects   138
Wine's Antimicrobial Effects   140
Wine, Lead, and Gout   140
Wine and Cancer   141
Wine and Medications   142

Wine's Nutritive Value   142
Fetal Alcohol Syndrome   143

Chapter 9. Visit to Gold Seal Winery   147

Ports: American and Portuguese   148
Sherries: American and Spanish   150
Sparkling Wines   155
Terminology on Champagne Labels   156
Production Techniques: Traditional, Transfer, Bulk   158
Champagne Corks   160

Chapter 10. Impromptu Meeting   165

Cork Structure and Properties   166
Cork-Borne Faults   169
Types of Oak: Structure and Properties   172
Toasting   173
In-Barrel Fermentation   175
Aging *Sur Lies*   176
Classification of German Wines   177
Botrytized Wines   180
Recioto Wines   182

Chapter 11. Visit to the Geneva Research Vineyard   185

Rootstocks   187
Phylloxera   188
Grapevine Structure   192
Training Systems   194
Organic Viticulture   199
Nutrition and Irrigation   200
Pest and Disease Control   202

Chapter 12. Pleasures of the Table   207

Origin of Western Views on Food and Wine Combination   208
Uses of Wine in Cooking   210
Concept of Flavor Balance   212
Influence of Flavor Principles   214

Examples of Principles in Action   218
Departure   222

Bibliography   **227**

Index   **237**

Preface

Conserve Water, Drink Wine improves wine enjoyment by providing useful information on how wines are produced. In addition, it gives practical suggestions on how to select, store, age, and savor wine.

The material is presented in the relaxed format of a discussion group. The reader becomes involved during their deliberations and visits to vineyards, wineries, and other sites in and around Ithaca, NY. Explanations are short, and questions arise naturally. The group evolves in the Finger Lakes Region of Upstate New York. Wine selection is presented by a local wineshop owner who explains how to choose wines. A columnist, a medical doctor, and a culinary lecturer present aspects of wine tasting, wine and health, and wine and food combination.

The text stresses aspects of grape culture and wine production that influence the quality, style, and aging potential of wine. In so doing, modern advances in grape cultivation and unique wine production techniques are covered.

Acknowledgments

It is with much pleasure that I express my appreciation to those who have generously given of their time to provide criticism of this work: notably Professor A. Rogosin, Dr. R. Smith, Mrs. Mae Jackson, and of course my wife. In addition, no work of this nature is the result of one person's labors. Generations of grape growers, winemakers, researchers, writers, chefs, and students have all contributed in their various ways to our knowledge of wine. Acknowledgment is thus given to the unsung multitude who have helped us reach our current "golden" age of wine. Finally, but not least, I appreciate the permission given by Marvin Myers to reproduce several of his perceptive cartoons.

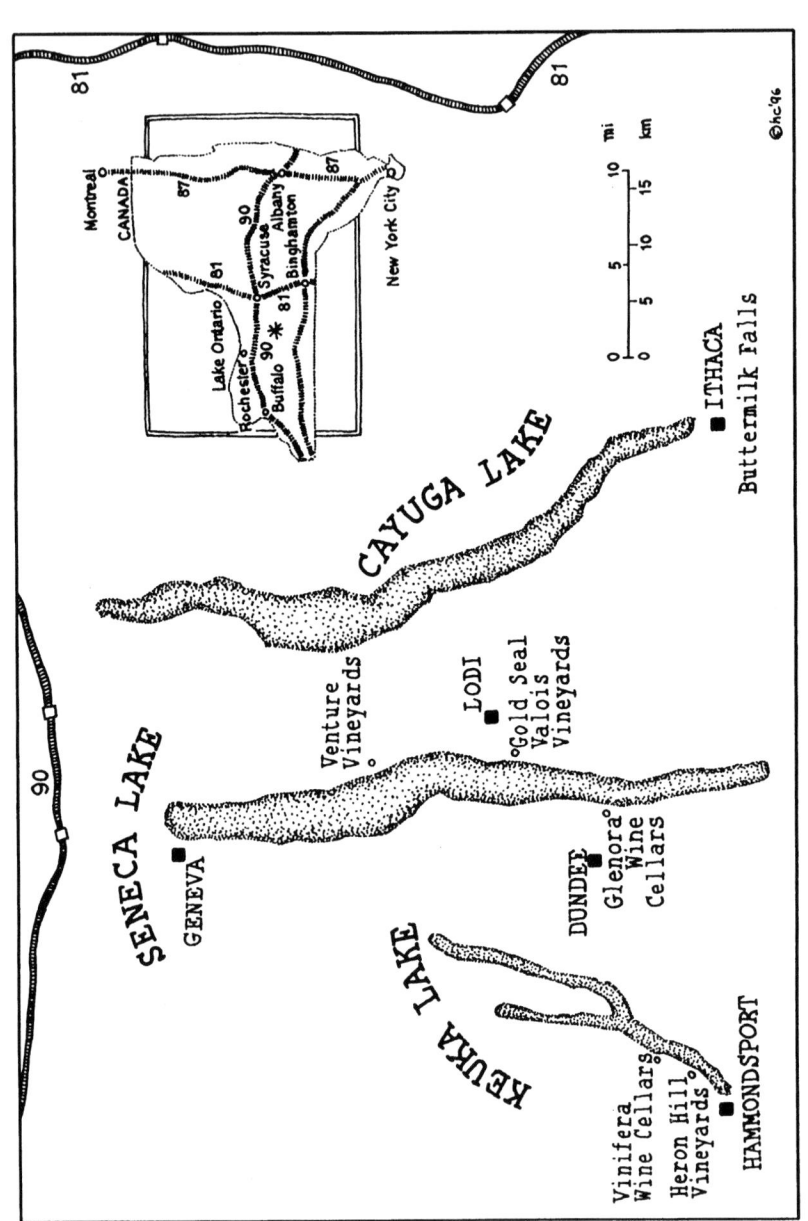

Map of Finger Lakes Region of Upstate New York

©hc'96

Chapter 1

Arrival in Ithaca

What a start to a sabbatical! Rain pelted the car and ricocheted off the road like bullets. Only a few hours ago the sun had been dancing off the St. Lawrence River when we left the cottage in Bic. The trip had taken longer than expected due to a protracted visit to Suzanne's brother in Terrebonne, north of Montréal. We'd been offered what was to be our dining room set–four aluminum folding lawn chairs, and an equally refined simulated-marble picnic table. Their addition required judicious unpacking and the removal of bicycles dangling at the back of the bulging Gran Torino. The wagon was jammed with all the essentials we could carry to stock our temporary abode in Ithaca, New York.

Nearing our destination after hours of night driving, we were becoming exhausted by the trip. The downpour had made the un-marked, newly paved asphalt especially slick. Suddenly, the road unexpectedly veered to the left, narrowed between sentinel-like barrels, and constricted onto a bridge. The road seemed to shrink even more as a megalithic transport lunged toward us from the opposite direction. Trying to play Sterling Moss under such conditions was more adventure than I wanted at that hour.

Nevertheless, within a few minutes we were turning right at the four corners in Dryden, a short distance from our destination. A few moments later we were glad to be backing the wagon into our as-signed spot, facing the patio of the apartment. A dash through sheets of water, and unseen but felt puddles, we reached the protection of the staircase overhang. Up the worn hemp-covered wooden risers and we were in front of apartment B204. A frantic search for keys, so carefully stowed for quick and easy access, and we were home.

It was not my fortune simply to roll out and inflate the air mat-tresses, which were to be our bed, and hit the sack. No! My wife is the living expression of the dictum that cleanliness is close to godliness. It took Suzanne less than a millisecond to spy dust, cobwebs, spots on the walls, and, worst of all, exoskeletal remains of insects. A glance at the kitchen was almost enough to make us vacate the premises immediately. The lateness of the hour, the foul weather, and the holiday weekend made our stay in B204 a necessity. Unfortunately, bloodshot eyes, a throbbing headache, and haggard looks were of no avail. I was dealing with a Ouellet! So, it was back to the car to rummage for our decrepit vacuum cleaner, a wash bucket, rags, and a can of Ajax. The can was the worse-for-wear due to a friendly nudge from our twelve-inch TV received at some stage during the trip. It was time to wash and sanitize! After several hours that included foaming Alka Seltzers, buckets of water, and surreptitiously dispatched centipedes (inhabiting the door seal of the disconnected fridge), it was time to blow up the beds. I can't remember when I fell asleep, but I assume it was between kneeling down to contact the convex squares of the mattress and achieving the reclined state. Anyway, the next thing I recollect was viewing sunlight in suspended dust as the rays streamed in through the uncurtained windows of our bedroom. Sabbatical had begun!

Whoever conceived of sabbaticals should be sainted. There is no more cherished privilege of college employment than those too infrequent years off. They breathe life into tired, disillusioned, harried teacher/researchers. The freedom to think, away from the daily shackles of lecturing, numbing committee meetings, grant applications, interim reports, and consulting demands, is close to bliss. This was my first sabbatical. During it I planned to rekindle my sagging research aspirations by tackling the genetics and physiology of the fungus *Botrytis*.

Botrytis causes innumerable plant diseases and is the bane of most grape growers. Nevertheless, under special weather conditions, the fungus inaugurates changes that can produce some of the most expensive, lusciously sweet, aromatic white wines. This nectar of the gods begins as juice from the wizened, purplish remains of once-healthy grapes. The first wines produced from such inauspicious beginnings were so amazingly exquisite that they were viewed as possessing magical, life-restoring properties. My interest in *Botrytis* was far more prosaic.

However, the day's agenda involved unpacking the car, getting groceries, and surveying our immediate surroundings. This became an almost daily ritual of walks, usually ending at either the Pyramid or Triphammer malls.

We were amazed that one could shop on Sunday, and on a holiday weekend at that! Such activities were *not* permitted at that time in Canada. Even better was the discovery of a wine shop just around the corner from the Food Mart, and directly facing the laundromat. I quickly realized what research I would pursue while Suzanne was guarding the laundry. I would return for the folding. I was appropriately housebroken.

My first day at work was quite a shock. The research collection of *Botrytis* mutants I had expected to find had succumbed to the ravenous appetite of mites. I would have to start from scratch. At least I did have a desk to myself, albeit in a room with four graduate students. This actually turned out to be a blessing. To be honest, the students were more interesting to talk to than the profs. Graduate students often have more time to chew the fat, so to speak. Also, they are less inhibited in what they say. University politics haven't yet infected and warped their perspective on life. Furthermore, it was here that I met Pat, a young, curly-haired, humorously irreverent PhD student. He exhibited the independence and droll disrespect for authority that I have always envied. He definitely expressed his Irish ancestry. We hit it off instantly and soon realized that we had a common love–wine. Along with myself, he became part of a group that played a critical role in my developing passion for and knowledge of wine.

Being on sabbatical, I had evenings free to pursue interests other than those directly related to my research. There was also time between preparing experimental media, reactivating fungal cultures, and planning experiments to do one of my favorite endeavors–spelunking in library stacks. The science library at Cornell University was housed in the same complex as the Plant Pathology Department. Thus, even short breaks in research activity could be profitably spent browsing among the multilayered caverns of Mann Library. The stacks were piled high with books of every size, color, and description, a paradise for any confirmed bookworm.

View of the Campus of Cornell University

It was on one of my early sorties to the stacks that I met Phil. He was perusing the culinary section when I accidentally bumped into him. I had to quickly sidestep to avoid being mowed down by the earnest librarian storming toward me with her load of scholarly tomes. In so doing, I knocked Phil's glasses off his rounded nose, and they landed precariously on the edge of one of the bookshelves. I profusely excused myself but he only laughed. "Another close encounter with Marian the Librarian? Some day we will find a pressed specimen of some hapless student caught off-guard by that menace of the book lanes. You must be new here if you haven't already experienced her single-minded dedication to her task."

"Yes, I'm here on my first sabbatical leave from Brandon University, a small institution some 130 miles west of Winnipeg, Manitoba."

"Canadian, eh?" Phil chuckled. "I thought I detected a bit of a British accent, but not enough to be from the UK, Australia, or South Africa. I've been up to Toronto once–a nice city–but no further. Doesn't Winnipeg have a hockey team?"

I'm always amazed when people detect an English accent in my speech. I may consider myself English, even though my ancestry is primarily Scottish.

"Yes, the Jets," I responded. "I hope you won't ask me anything about them. Although I'd be ostracized for saying so, I have no interest in what many Canadians regard as their national sport."

"Fear not!" said Phil, waving his hand expressively. "I'm not much for sports myself. Teaching and research take up all my time. I try and avoid committee work like the plague."

"I know the feeling!" I replied. "I promised myself that on sabbatical I'd leave the evenings free for interest reading and enjoying some time with my wife. Because she doesn't have a *green card*, she can't even supply teach. In a way, though, it's nice for her to take a year off from teaching. For once, she'll be free to go for walks, read, get back to painting, and simply enjoy life for a while.

"You were looking through the section on food science just before my collision with you. Do you teach in that field?"

"An aspect of food science, yes," returned Phil. "I'm in the Hotel School, not far from here. It is just across from Day Hall. I teach courses in the principles of menu design, food presentation, and beverage service."

"Then it seems that I've bumped into the right person," I joked. "I'm searching for a useful book on food and wine association. Do you know of any?"

After a few moments of reflection, Phil replied. "Regrettably no. There are plenty of books supposedly on that subject. However, they typically consist of a list of the author's personal wine preferences and food recipes. The suggested pairings are a source of frustration as they are rarely obtainable by the reader. Worse, though, is the impression that there is some perfect association between specific wines and food. To me, it's equivalent to suggesting that particular spices are essential to enjoying certain foods. Why should people expect things to be different with wine and food associations? Nevertheless, on a different matter, if you are interested in wine, there is a Bacchus Society here in town."

"What is it, a wine tasting society?"

"Yes. The group meets once a month to sample wines. It's more socially inclined than academic, but does allow people to try a broad range of wines inexpensively. If you would like, I'll contact the treasurer of the society about your joining?"

"That'd be super! My name is Ron Jackson, by the way. I can be reached most easily through the secretary in Plant Pathology during the day. Being members of the Society would give Suzanne and me a chance to meet people with similar interests. I'm planning to audit a course on viticulture given by Bob Pool from the Geneva Research Station. Would it be equally possible to sit in on your beverage course?"

"It would be a pleasure to have you, but the course currently is full. In addition, it's the policy of the Hotel School that neither regular nor visiting faculty may audit our courses gratis."

"Ouch!" I exclaimed. "At the price of courses here, and on a sabbatical salary converted into American funds, that does make a difference!"

"You may be interested in a small self-study group that a few of us from the Bacchus Society have been contemplating forming," responded Phil. "Although I teach the beverage course in the Hotel School, and enjoy wine immensely, my professional training is in menu preparation, not enology or grape growing. It would be a learning experience for me as well. Would you be interested in becoming a member?"

"It sounds great. I've been tinkering with the idea of giving a wine course when I get back to Brandon. Such a group would help focus my tentative preparation. Also, I know of a student in Plant Pathology who might also be interested in participating in such a group."

"I'll contact those who were interested in the group and get back to you," said Phil, as he picked up his glasses, still precariously positioned on the bookshelf where they had landed.

It wasn't for a couple of days that I had a chance to talk with Pat. I was at my desk trying to identify fungi collected under wild grapevines when he walked in. He had been away doing field work on *Botrytis* onion leaf spot. Although less interested in the details of winemaking than I, Pat thought the group would provide a great opportunity to expand his wine knowledge. He was also interested to hear about the Bacchus Society. Although it was Pat's second year at Cornell, he hadn't heard of the Society's existence. He knew his wife would be interested in joining.

The next stage in my association with the group happened when Suzanne and I were out cycling on the weekend. Because of the cost of gas and the thirst of the Torino, we used the station wagon as little as possible. Our original plan had been to follow the bus route to the university. The route passed directly in front of the University Park Apartments, where we were living, and had a stop just down Tower Road from Bailey Hall, where I worked. It passed some of the most gracious homes. On cycle, we could stop and study more closely these handsome structures and tour the university campus.

On our way we took a short cut down Brandywine Road. To my surprise I saw Phil cutting the grass in front of one of the houses. We pulled over so that I could introduce Suzanne. My expression must have communicated my surprise that he lived so close. I subsequently learned that he had recently taken possession of the house. He invited us in for a bite, but we explained that we had just eaten and were out for the exercise. We wanted to enjoy the nice weather while it was holding. Phil asked if we had been to the Commons yet. We admitted our ignorance about the Commons.

Phil explained that it was part of the downtown where several streets had been closed to traffic and turned over to pedestrian use. If we liked Blue Grass music, we'd probably be interested in a free concert being given today. There often were musical performances, craft shows, art exhibits, and other events on Saturday afternoons. It sounded like fun, so we decided to go to the Commons instead. He also suggested that we drop in on the Clever Hans Bakery for fresh homemade bread. He guaranteed that we would not be disappointed. As we were preparing to mount our trusty bikes, Phil asked if I was still interested in participating in the small wine-study group he'd mentioned at the library. Not only was I interested, but Pat in the department also wished to participate. Phil asked if Suzanne wanted to be a member as well. She appreciated the invitation but declined. She enjoyed wine with food but had little desire to learn about how grapes were grown or wines made. Suzanne preferred to use her time reading and painting. Phil wondered if Pat and I would be free to come to a formative meeting this coming Tuesday evening about 7:30. I knew I would and promised to get back about Pat. With that good news we set off, this time for the Commons.

Because of the steep slope down to the main part of town, we decided to weave our way through the meandering side streets, rather than risk a pell-mell dash down Court or Buffalo streets. We were glad we did as we happened to pass through a collection of splendid residences snuggled among the woods that clothed the eastern slopes of Cayuga Lake. We were astounded at the sensitivity with which the homes had been incorporated into such a natural and idyllic setting. Almost every property had its own brook trickling its way through the site. It was as if we had stumbled onto a fairyland. It made our little house on the prairie back in Brandon look pretty humble.

One of the foremost pleasures of living in a college town, especially in wine country, is the number of fine wine stores. Checking out their selections became one of my favorite pastimes. In addition, they had knowledgeable staff, willing and able to make sound suggestions and answer questions. What a change from the way business used to be conducted in Canada's provincially run liquor control stores. The notion of *control* was once so strong that employees were not permitted to make suggestions. Knowledge of the product, if not actually discouraged, was not promoted.[1]

It was in one of the wine stores in the Commons that we first encountered printed descriptions of wine attached to the holding bins. The comments found on file cards introduced us to the wines of Rioja. Informative notes also induced me to try my first Barolos from Italy. In addition, it was here that we first met Peter Phillips. Later, he met with our group to share and discuss wine selection.

Although on severely restricted funds, Suzanne and I maintained our policy of giving each other personal money. Most purchases

"Here's a jug of wine and a loaf of bread. . . . Do you need a thou?"

---

[1]Thankfully, conditions in Canada have improved, and in some ways they now surpass what so enthralled me in Ithaca's wine stores.

came out of a joint account into which both our salaries went. Out of this we paid ourselves an allowance of fifty dollars a month. With this we could purchase items without getting joint approval. Most of Suzanne's personal money disappeared into a Kip Sanderson painting of a weather-seasoned fisherman. She had seen it at the same store on Dryden Road where we had bought our Boston rocker. Whereas most of our furniture was modern modular (empty wine boxes obtained from the Triphammer Mall), we needed something more sturdy and comfortable for reading. Her acquisition of the fisherman has survived longer than mine. Only two bottles of wine remain out of what I purchased with my allowance money. The 1975 Château Beychevelle is scheduled to be consumed in the year 2000, on our twenty-fifth wedding anniversary. I have not decided on what auspicious occasion the 1971 Cinque Castelli Spanna by Antonio Vallana should be opened. Suzanne's fisherman still stands guard over the fireplace at our cottage.

In contrast with the hurried manner in which I now select wines, I had the time then to check out several stores, compare producers, vintages, and prices before placing down my greenbacks. I still marvel at the incredible range and quality of wine I could purchase on such limited wherewithal.

Chapter 2

The CWDW Society

On Monday night I phoned Phil to tell him that tomorrow evening was fine for Pat and myself. I asked if we should bring anything of a liquid nature. He replied that the others thought it would be fun if we each brought a bottle of our favorite wine. Dave, a chartered accountant, had also wondered if Phil would give an introductory talk on the origin of wine to start us off. Because he was currently discussing this topic in his beverage course, Phil agreed. It would also provide an example of how we might conduct future meetings. One or more of us could research a topic and give a presentation to the group. This seemed a good way to proceed as it would spread the work around. Also, I knew from experience that the best way to learn anything is to teach it. I'll never forget the year I taught upper-level Physics in high school. It was the same course that I had resolutely avoided taking when I was a student. Necessity is not only the mother of invention!

After a lovely weekend and beginning to the week, Tuesday turned out to be overcast and drizzly. By the time the bus dropped me off at the apartment in the late afternoon, the drizzle had settled into a downpour. I soon came to dislike rainy spells for reasons other than having to sport an umbrella. The apartments were built on the side of a hill sloping toward Cayuga Lake. Although the decline helped carry off heavy rains, downpours also tended to wash populations of drowning earthworms onto the pavement. The flattened remains of these hapless creatures generated a distinct odor. Although bearing no direct relationship to wine, I began to

---

Pat had jokingly dubbed the group the CWDW (Conserve Water, Drink Wine) Society after spying the bumper sticker I had on the station wagon. The sticker had been produced in California during one of their periodic droughts.

recognize the same essence in some wines. Only years later did I realize that I was hypersensitive to a trace contaminant formed in the presence of diacetyl. Diacetyl is associated with the buttery character of several wines. Thus, when wine critics crow about the rich buttery flavors of wines, I cringe. I'm afraid of detecting night crawler essence.

Because of the inclement weather, I took the station wagon. Phil's house was a two-story, frame, colonial-revival house. It was painted a soft yellow, which appeared off-white in the gloom, and was set off by chocolate brown shutters and trim. The imposing solid-wood front door was elegantly framed by sidelights of frosted glass exhibiting an ivy leaf design. Phil's diminutive wife Gloria met me at the door and ushered me into the large central hallway. At its end stood a stately antique grandmother clock. A glance to the right revealed an elongated, cherry dining room set that would have made Suzanne swoon. We had longingly admired a similar set in the store where Suzanne had purchased her oil painting. It is part of the Canadian psyche to secretly covet the quality choices our American cousins take for granted. However, Gloria directed me to the left, which led into the parlor.

Phil was talking with a tall, slim, attractive fellow with strikingly azure eyes. Even I couldn't miss his blue eyes. Suzanne was always commenting on people's eye color, while I was almost oblivious to this human trait. Phil turned and introduced me to Dave.

"You'll want to talk to Dave," Phil began. "He's the treasurer of the Bacchus Society. I forgot to mention when I phoned him several days ago to contact you about joining the Society."

Dave's pinstriped navy suit seemed appropriate to his being a senior partner in one of the local accounting firms. He was about six feet tall. His auburn hair was streaked with silver and was beginning to thin on top. His warm humble nature made it immediately apparent that I would feel comfortable in his presence.

The next person introduced was Antonio, or Tony, as everyone except his wife called him. Tony was a history teacher whose love of winemaking probably would have turned him into a vintner, had he not grown up on the Lower East Side of Manhattan. Phil had met him, like most of the group, at the Bacchus Society. Although a second generation Italian American, Tony's mannerisms and accent quickly denoted his ethnic origin. His complexion and stalwart frame further enhanced his Italian appearance. The whole family still spoke Italian at home and made frequent trips back to the old country.

My next acquaintance was Claude, an expatriate Frenchman who had emigrated to the United States several years ago. He had come to the United States with his wife, whom he had met when she was on a teacher exchange program in Strasbourg. Claude had beet-red hair and freckles. His robust stance and appearance made me think of Eric the Red. He was a radiologist at the Tompkins County Hospital, almost directly across the lake from where we were.

In the interim, Pat had arrived. So, it was now my turn to do the introductions. After some additional chitchat, Phil invited us to sit in the dining room for our first deliberation.

ORIGIN OF WINEMAKING

"Because of the impromptu nature of the meeting," Phil began, "I've agreed to be our first speaker. As it is usually best to begin at the beginning, I'll briefly review what we presently know about the origins of winemaking. As with most food fermentations, such as beer, mead, bread, and cheese, the discovery of wine production is shrouded in mystery. All these fermentation processes predate recorded history.

"Because grapes so readily undergo fermentation, one cannot logically envision a single time or place where winemaking began. It's more reasonable to suppose that the intoxicating effect of fermented grapes occurred frequently before the realization that crushing promotes this transformation. If such accidentally fermented juice is called *wine*, then winemaking undoubtedly occurred wherever humans stored grapes in vessels. However, if we define wine more restrictively as the clear, stable product of fermented grape juice, then human society had to have passed to an agrarian culture producing sealed vessels. Airtight, relatively impervious containers are required to protect wine from exposure to oxygen. Without this protection, wine soon becomes vinegary and undrinkable.

"As far as we are presently aware, deliberate winemaking evolved somewhere in the Caucasian region. This includes portions of modern-day northern Turkey and Iran, as well as Armenia and Georgia. This may reflect the geographic site where agricultural peoples first overlapped the southern distribution of wild grapevines. However, this does not explain why winemaking did not develop where agriculture extended into the distribution of other grapevine species, such as in China or North and Central America. What may explain this anomaly is the distribution of the progenitor of the wine yeast.

"Different strains of this yeast are currently distributed worldwide. They are variously called the wine, brewers, or bakers yeast. The production of wine, beer, and leavened bread all first occurred in regions in or adjacent to the Near East. Linked with human activity, this yeast is rarely found in natural environments. The natural habitat of the yeast, called *Saccharomyces cerevisiae*, may be the gummy sap exudate of oak trees. Interestingly, the southern distribution of oaks in Eurasia almost perfectly matches that of wild European grapevines (*Vitis vinifera*), commonly termed the wine grape."

"May I interrupt for a moment?" Tony said. "You mentioned earlier that grapes ferment easily. Why, then, do I add yeasts when I ferment grape juice?"

"You normally add a strain of wine yeast to speed the process of fermentation. This gives less time for spoilage yeasts and bacteria to generate undesirable odors. Wine yeasts are also added because most other yeasts are susceptible to their own alcoholic waste products. Thus, in the absence of wine yeasts, grape sugars may not completely ferment. This leaves the wine not only with a sweet finish, but also particularly susceptible to microbial spoilage.

"One of the more fascinating features about grapes that I haven't mentioned thus far," continued Phil, "is their phenomenally high sugar content. Wine grapes commonly reach sugar concentrations of 20 to 25 percent. No other fruit has this property. Most other fruits remain relatively starchy, and if they produce juice, it is considerably less sweet. Wine grapes are almost concentrated sacs of sugar. The next time you happen to be out in a vineyard, taste some fully ripened wine grape. You'll experience what I mean!"

"I can certainly confirm that," asserted Dave. "However, earlier in the season, the fruit is very tart. What happens to the acidity?"

"Some of the acids are degraded by enzymes in the fruit as they mature. This is especially so with malic acid, one of the two main acids in grapes. The other acid, called tartaric acid, does not break down as readily. In addition, tartaric acid is resistant to the action of most microbes. This is essential to the grape's retaining considerable acidity up to harvest time."

"You indicate that acidity is important," responded Claude. "Of what conceivable value is acidity? I like my wines smooth, not sour."

"The retention of acidity is vital to the quality of wine for several reasons. Acidity helps prevent the growth of most spoilage yeasts and all food-poisoning bacteria. Acidity is also essential to maintaining the color of anthocyanins, the red pigments in grapes. In addition, acidity gives wine its lively, fresh, clean taste. Without it, wines would taste flat."

"Could we return to the origin of wine for a moment?" I said. "I found your arguments on the origin of winemaking in the Caucasus intriguing. Nevertheless, is there not additional evidence supporting this view?"

"Oh my, yes!" declared Phil. "Most recently, I've been fascinated by arguments of linguists. They have noted some illuminating differences within Indo-European languages for the words *wine*, *grape*, and *vine*. For example, the term *grape* differs considerably among these languages. This implies that ancestral Europeans had a knowledge of, and a local name for, grapes before Indo-European speakers migrated into Europe. In contrast, the similarity between words for *wine* in most Indo-European languages suggests that the knowledge of winemaking had a focal source. As the use of wine spread, so did the term used for the product. It is interesting to note that linguists locate the origin of Indo-European languages in the Caucasus.

"Another line of argument is provided by historians. For example, most Eastern Mediterranean myths locate the origin of winemaking to the east, along the edges of the then-known world (roughly the Caucasus).

"Archaeologists have unearthed wine-related artifacts," continued Phil, "but these are all too recent and advanced to suggest a site for wine's origin. For example, the oldest known wine residues come from northern Iran, and date about 4000 B.C. The oldest drawings that clearly show wine production are of wine presses found in Egyptian tombs, dating back some five millennia. Because grapes do not grow indigenously in Egypt, the grapevines depicted probably originated from cuttings or seeds imported from northern Syria."

DOMESTICATION OF THE GRAPEVINE

Tony interjected, "I've read that the cultivars found in Italy were spread by early Phoenician and later Greek colonizers. Does this view have any historic or scientific backing?"

"Except for some excerpts referring to Greek colonists in southern Italy, the written record is silent. As far as botanical remains, the evidence is insufficient to either confirm or deny this possibility. The best evidence I have seen relates to changes in the structure of pollen as vines became domesticated. During domestication, vines possessing bisexual flowers became frequent. This development is associated with the reduced occurrence of sterile pollen in sedimentary deposits throughout Europe. This shift is detectable, though, long before the movement of Romans into Celtic Europe, or the rise of the Phoenicians. One breeder I know thinks that if domesticated wine grapes were carried west and north, their influence may remain only in traits picked up by crossing with local vines. In most cases, local grapevines are better adapted to the prevailing soil and climate conditions than imported varieties. This view is also held by several Russian and French researchers."

"I'm fascinated that domestication was linked with an increase in the frequency of vine bisexuality. What is the situation in wild grapevines?" asked Pat.

"The usual situation in wild grapevines is for the plants to be either male or female. That is, the flowers of individual vines produce only functional male or female parts, even though the sexual structures of both sexes may be present. Nevertheless, some grapevines show both functional female and male parts. The functionally bisexual flowers are self-fertile. It is assumed that during cultivation, unfruitful male vines were uprooted. With the male vines removed, the remaining unisexual female vines would also have become unproductive. This would have resulted in a natural selection for fruit-bearing, bisexual vines. With a reduction in wild female vines, linked with deforestation and the expansion of agriculture, the production of sterile pollen, and its deposition in sediments, would have declined."

"What about the origin of our own American cultivars?" questioned Dave.

"Although American cultivars are of comparatively recent origin, their precise derivation is often unclear. For example, the cultivars 'Concord,' 'Catawba,' and 'Niagara' are variously considered to be either direct selections from wild strains of the fox grape (*Vitis labrusca*), or accidental hybrids between this grape and the European wine grape (*V. vinifera*). The latter view was espoused by the late Dr. L.H. Bailey, Cornell's most famous researcher-horticulturalist. It is in his honor that Bailey Hall on the campus was named.

"More recent cultivars, such as 'Dutchess' and 'Delaware,' are derived from intentional crossing between the fox and wine grapevines. Further south, cultivars such as 'Noah,' 'Lenoir' and 'Norton' are simple or complex hybrids involving two or more of the fox, summer, sweet winter, or wine grapes. Still further south are cultivars based on the muscadine grape, for example 'Scuppernong' and 'Noble.'

"Although some of these cultivars are still widely planted, the new kids on the block are classic European varieties and the so-called French-American hybrids. The latter were developed in France to combine the disease and pest resistance of American grape species with the winemaking characteristics of European grapevines. Their work was so successful that the new cultivars began to supplant traditional varieties. In the more famous French wine-producing regions, this came to be viewed as a threat to the continued dominance of traditional wines. To protect their position, producers from the latter regions convinced French legislators to pass laws strongly restricting the planting of French-American hybrids. The effect has been a reduction in the coverage of these hybrids from about 30 percent in 1955 to less than 5 percent today.

"While the planting of French-American hybrids has declined in France, and elsewhere in Europe, their use in North America has risen. Nevertheless, there is even more interest in cultivating pure *vinifera* varieties such as 'Chardonnay,' 'Riesling,' 'Cabernet Sauvignon' and 'Pinot noir.' Personally, I regret this trend because European cultivars do not produce more distinctive wines here than elsewhere. Do we really need more Chardonnay and Cabernet wines? I would prefer to see our region specializing in the distinctive attributes of French-American hybrids and other new varietals. This would give our wines a unique market niche in the world.

"Normally at about this point in my lecture I mention statistics on grape and wine production, to show the regional and global significance of grapevines. However, I feel statistics are not what you want to hear. However, you might be surprised to know that grapes are the world's leading fleshy fruit crop, surpassing the production of oranges, bananas, and apples. In addition, I hand out a sheet[1] listing books, journals, and magazines that provide information on grape and wine production.

"If you want precise information, books are probably your best bet. Scientific journals are fine, but are written for other scientists. I look at some of the articles and wonder how they relate to grape or wine production. Wine magazines are great for color photographs, travelogue information, human interest stories, glowing reports on tastings of rare wines, and personal opinions about particular wines. As you can gather, these are not particularly useful sources of objective information. Food and wine magazines are of a similar genre, except that they occasionally possess a worthwhile recipe or two.

"As Mark Twain used to relate about halfway through his speaking engagements: 'It's a terrible death to be talked to death.' Thus, I feel it's time for a break."

WINE SAMPLING AND TERMINOLOGY

As if on cue, Gloria appeared with a tray laden with an incredible assortment of hors d'oeuvres. It became obvious to those of us who did not already know, that Phil and his elegant wife knew how to prepare sustenance that appealed not only to the taste buds but also to the visual senses.

Mouthwatering as our break was, we were soon back to the serious business of wine tasting. It was now time to relish the

---

[1]See Bibliography for a modern version.

samples each of us had brought. As we arranged the glasses we had brought in front of us on the burnished cherrywood table, Phil opened a bottle of Montrose Estate Mudgee Chardonnay.

"While you pour the wine," he said, clearing his throat, "I'll tell you where the wine came from, and why I chose it. The wine comes from a relatively small winery in the Mudgee region of New South Wales, on the eastern coast of Australia. There are several reasons, other than its fine quality, to justify my presenting it. Its rich yellowish golden color appeals to the eye and the melon/peach fragrance and harmonious flavors are a joy in the mouth. Its lingering flavor in the mouth (finish) establishes its regal stature. It is incredible that the wine is available at the price of nondescript inexpensive wines! Although I love the wine, it is intriguing to realize the conditions under which it was produced. It originates from a latitude that would be the equivalent of Morocco and Lebanon in the Northern Hemisphere. New South Wales is ostensibly subtropical, and based on European concepts, the region would be ruled out as a site for the cultivation of 'Chardonnay' grapes. In Europe, 'Chardonnay' is typically grown only in cool regions. Admittedly, the wine comes from valley slopes averaging about 1,800 to 1,900 feet above sea level. This provides cooling and gives the vines some protection from occasionally severe spring frosts. The wine, or more precisely the Mudgee region, is unique as far as I know in being the only Australian region to have its own appellation control system. Together, the producers certify their wines as coming exclusively from within the Mudgee region."

Although more difficult to assess the quality of a wine in which the presenter is so enthralled, there was no doubting the superb attributes of the wine. The group genuinely agreed on its quality and was amazed that such a flavorsome Chardonnay could come from a warm climate.

"One final note" said Phil. "Could we avoid the term *nose* in referring to a wine's fragrance? It is one of my pet peeves. Although legitimate to some because it comes from French, to me *nose* refers to a part of the human anatomy. More significant, though, is the term's association with snobbism. I strenuously object to wine being considered elitist. Wine is a food beverage, designed to be taken either with, shortly before, or after a meal. In addition, there is no

"I think I'm getting closer! It's either a red or a white."

need to employ translated foreign terms when English already possesses sufficiently descriptive words. Fragrance adequately applies to all wine-related aromatic sensations, except off-odors. If one needs greater precision, there are the terms *aroma* and *bouquet*, which refer to grape and fermentation/aging odors, respectively. Well! Now that I've gotten that off my chest, I'll stop."

Dave was the next to volunteer his selection for scrutiny, and he conscientiously avoided the term *nose*. "I have brought what I believe is an excellent quality Californian Cabernet Sauvignon. It is a Heitz Napa Valley Cabernet. This is not one of Joe Heitz's famous single vineyard bottlings, like Martha's Vineyard or Bella Oaks. Nevertheless, I believe you will find the wine has a captivatingly rich, fruity, black current aroma. This is combined with a complex cigar-box bouquet. In the mouth, the wine is smooth, without harshness, but with enough astringency to permit it to mesh with a meat dish. The wine also shows a long finish, one of the best indicators

I've found of a great wine. The wine proves that a winemaker with finesse can combine several wines from different sites to enhance their best qualities. Blends are often better than any of the individual wines alone. Such wines may not have the status appeal of estate-bottled wines, but can be of better quality and less expensive!"

With the first two wines of such fine quality, I began to feel edgy about my selection. I had had neither the time nor the funds to buy two bottles–one to try in advance and one for the tasting. I hoped that the proprietor of East Hill Liquors knew his stuff. It was on his recommendation that I'd purchased the Riesling. Worse still was that mine would be the last sampled, as it was sweet. Anyway, it was too late to make any changes. I should just enjoy what was coming and hope for the best.

Pat may have had the same feeling of trepidation, but he decided to put caution to the wind and agreed to be next.

"Although I don't know that much about my selection," admitted Pat, "I hope you'll find it worth sampling. It's one of my favorite wines, not only because of its moderate price, but also because the label is amazingly modest."

The label possessed only austere black printing on what appeared to be unbleached newsprint. It was a label devoid of all modern marketing concepts. The wine was a Vinho Tinto Garrafeira from Carvahlo, Ribeiro, and Ferreira, Portugal.

Pat continued. "What grape varieties were used in its production I don't know, nor could I detect from what region the wine comes. Regardless, the wine possesses one of the features I especially appreciate in a wine–a slow opening. Unlike the two previous wines, the Vinho Tinto initially shows little bouquet. However, as you continue to swirl the wine in your glass, an almost seraphic fragrance begins to arise from the wine. This is the one wine I have had that blossoms like a flower under my nose. Although the opening of a wine's fragrance is often described in reports on wine tastings, I've seldom detected it myself."

From the looks of pleasure, everyone seemed to be experiencing the flowering of the bouquet. Dave conceded that he would not have thought that a wine this old would still have such a young color. At fifteen years of age, it was only showing a slightly brickish overtone. Phil asked Pat about the meaning of the expression *garra-*

feira on the label. Pat stated that he didn't know what it signified. Phil then explained that *garrafeira* was a term used by Portuguese firms to designate their finest table wines. Often these are blends of wines produced in different regions of the country, and probably from several grape varieties. Under most circumstances, such as this, garrafeira wines are excellent buys. Pat beamed with pleasure that his selection had met with such broad approval.

The next to put forward his selection was Claude. "What I brought along is one of the few remaining bottles I have of an Adelsheimer Pinot noir. It comes from the Willamette Valley in Oregon. I was lucky enough to be there a couple of years ago on vacation. I was amazed to find so many Pinot noir wines. Even more startling was their excellent quality and moderate price. They reminded me of the better Burgundies I'd tried back in France. Apparently the conditions in Oregon are almost a match for those in Burgundy, except that Oregon's climate is less subject to yearly variability. David Adelsheimer, from whose estate the wine comes, noted that most producers have astutely followed the winemaking practices of Burgundy. Regrettably, I didn't have more space in the car at that time to bring back more wine. Here's hoping this bottle is as good as the others I've opened!"

Claude need not have worried. My view was that if Burgundy could produce wines like this then their reputation was well-deserved. But I soon gathered from the discussion that there was a wide diversity of opinion on Burgundian wines. Dave emphasized that he loved Pinot noir wines, but had become disillusioned with most Burgundies. While the top producers generally produced consistently good wines, they were prohibitively expensive. A tasting with the Bacchus Society several years ago had uncovered several nondescript wines at over thirty-five dollars a bottle. Phil stated that some important négociant/éleveurs in Burgundy were purchasing property in Oregon. He had personally tried a Drouhin Estate Pinot noir which was exquisite.

Regardless of the quality of red Burgundies, the Pinot noir we had from Oregon was luscious. The ruby red color was almost as much a pleasure to the eye as was the flavor to the palate. We lingered over the wine for some time before progressing to Tony's choice.

"I trust you will forgive me for being a bit parochial," Tony grinned. "However, I have an Italian wine that I'm certain none of you have tried before. It comes from southern Italy, not from my region around Naples, but to the east from the "heel" of Italy, Apulia. The region contains many local cultivars little known even outside Apulia, let alone Italy. When vinified with a skill that has grown by leaps and bounds recently, these varieties can produce flavorful, aromatically distinctive wines. The varieties involved in my selection, a Notarpanaro from Cosimo Taurino near Salento, are the 'Negro amaro' and 'Malvasia nera' grapes. The wine has a deep, rich, red, almost black color. In its nose–whoops! I mean *fragrance*–I find a sharp spicy aspect associated with an opulent jammy aroma. At times I almost think I detect some chocolate overtones. But for me, the greatest thrill is in the mouth. It's a meaty, tangy wine, but not inky. It is a wine that stands on its own, no wimpish imitation of some other regional wine."

With an introduction like that, how could anyone not concur? But it was not for lack of backbone that people agreed with Tony. The wine was indeed splendid and unique, as he said. It made me wonder how many other potentially remarkable grape varieties were languishing almost unknown in their country of origin. Phil voiced a similar sentiment when he wished out loud that winemakers would be more adventuresome in the cultivars they planted. However, he acknowledged that it was risky for winemakers to follow his suggestion. There was no sense in producing wines people would be apprehensive about buying.

"The majority of wine consumers are a conservative lot," Phil said, "either due to timidity, based on a lack of knowledge, or due to prejudices picked up from others more vocal but no more knowledgeable. If other cultivars were given the same fastidious care as classic varieties, what an increased and glorious range of superb wines we might choose from!"

"Well, I guess it is my turn now, as the presenter of a dessert wine," I said. "I must admit that I don't know how the wine will show. I have not had a chance to try it myself. I was looking for a spätlese from the Rheingau when the store owner recommended the wine you have now. Because it was from a grape variety and region I knew, I did what most consumers do in a pinch–stick with what you recognize. As a Riesling it should show a rose/pinelike fragrance. If affected by noble rot, it should also show elements of apricot–or as my wife comments, the aroma of pumpkin cooked in brown sugar."

"I hear that you work with the organism that causes noble rot. What exactly is it?" inquired Claude.

"It's a fungal infection that, under special climatic conditions, promotes the drying and concentration of the flavors in the fruit. It also modifies the varietal aroma and adds its own special fragrance. However, under more usual conditions, the fungus causes a bunch rot that makes the grapes useless for making wine . . . or anything else."

While I was pouring the wine and showing the label, Phil asked if I knew the meaning of *grünlack* on the label. Upon looking more closely at the label I noticed *grünlack* in the position where you would expect to find the name of the grape variety, if given, on a German wine label. I admitted that I did not know the meaning of the term, nor did I know of a variety so called. Although the wine was supposed to be a Riesling, the varietal name was noted nowhere on the label.

"I'm not trying to pretend great knowledge of German wines," interjected Phil. "However, I'm lucky enough to have visited the Schloss Johannisberg estate from which the wine comes. You will note that the label displays a coat of arms and has a green background. *Grünlack* in German means green label; this signifies the upper grade of *spätlese* produced by the Schloss Johannisberg estate. The marginally lower quality, dryer, *Weisslack* (white label) *spätlese* has a white background and a representation of the *schloss* (castle) on the label. Current German law does not allow vineyards to distinguish between subgrades of the six *prädikat* wine categories. Thus, some top producers have developed color-coded labels (Schloss Johannisberg) or color-coded capsules (Schloss Vollrads) to notify their own distinctive subcategory styles. Several top producers also do not note the variety on their labels. They expect their customers to know that they only use 'Riesling,' not any lesser grape variety. In addition, several estates have been permitted to retain the old custom of designating the wine only by the vineyard name, for example Schloss Johannisberg. Normally, the town in whose jurisdiction the vineyard or group of vineyards is located must appear on estate-bottled wines."

By this time everyone was savoring the wine except Phil and me. The expressions of pleasure, combined with serious concentration, indicated that my offering was meeting with approval, and possibly admiration. When I raised my glass to my nose, I realized why. It was mighty good! I love German wine and have had some great wines from that enchanted land, but this was special! The interplay of sweetness and acidity was unbelievable–a balancing act fit for the high wire. It was clear that I had to pay a return visit to East Hill Liquors to purchase more and thank the owner for his excellent recommendation. Some of this wine was going back with me to Brandon.

Phil was the first to break the thick silence that had come over us. "If we have meetings like this very often we will not be able to accept good but ordinary wine."

"Were it only possible to live on such an elevated vinous plain all the time," said Pat.

"From the looks of it, I assume that everyone has agreed that our little group should continue?" Phil asked jokingly.

"No doubt about that!" proclaimed Dave.

"Does anyone have a suggestion for a theme for our next get-together?" Phil asked.

"Since the grape harvest is rapidly approaching, why not go to a 'pick-your-own' vineyard, collect grapes, and make some wine," suggested Tony. "We could make it at my place. I make wine every year and have all the equipment necessary."

"I like that idea," I said. "I've read and talked about winemaking in my Economic Botany course, but I've never taken the risk of making homemade wine."

As the group liked the suggestion, we decided to arrange a trip to Venture Vineyards. Tony often went there to harvest grapes for winemaking. He volunteered to contact Phil when the grapes would be ready for picking. Phil would then contact the rest of us.

When I left that evening, it was still raining outside. Inside, however, there was a warm radiance emanating from all those sun-ripened, fermented grapes I had so pleasurably downed. I was glad it wasn't too far to drive back to the apartment!

"I think it's time we discussed your wine appreciation course."

Chapter 3

Visit to Venture Vineyards and Home Winemaking

The sky was a soft autumnal blue, dotted with puffy cherublike clouds winging their leisurely way across the heavens. The sun's warmth gave an impression similar to that of snuggling up to a blazing fire in winter. It was one of those beatific days when travel photographers must take their fall shots. It was a day to live, especially realizing what was soon to come. Our small group could not have chosen a more perfect occasion to go vineyarding at harvest time. Klaus Muller, a graduate student in viticulture at the Geneva Research Station, had agreed to come along and answer questions we might have.

I was up early for a Saturday morning–6:15. As it was still chilly, I put an old sweater over my *Mycology is Better than Yours* T-shirt. Harvesting grapes in the afternoon would be no dress-up affair. We had agreed to meet at Venture Vineyards. It was owned by Mel Nass, a member of the Bacchus Society.

Venture Vineyards is located on Upper Lake Road, just north of the junction of Highways 96 and 414. The vineyard contained both conventionally harvested and pick-your-own sections. Mel, a former senior executive with IBM in Rochester, had fallen in love with fruit growing and had "retired" to the life of a grape grower.

When I drove in shortly after nine, most of the members were milling around the courtyard outside the sales office. It was a weather-beaten wood frame building used for signing up for grape picking and as a retail store. There, you could purchase a variety of locally produced preserves and regional art and craft items. Mel was taking the opportunity to question Klaus about some recent work being conducted at Geneva on cold hardiness.

"What's the latest news on chemical protectants against frost damage?"

"As you probably know, we are studying the effects of several cryoprotectants at the research station," responded Klaus. "Of these, we have found DuPont Surfactant WK to be the most effective. However, to have maximal value, it must be applied several days before frost conditions develop. If you would like, I'll send you a copy of our article published in the *American Journal of Enology and Viticulture*."

"I'd appreciate that," commented Mel. "We had more frost damage than expected this past spring. It caused considerable base-bud activation and the expense of late pruning."

At this point, Pat came roaring in on his flaming red Suzuki. As our complement was now full, we headed off into the vineyard.

INFLUENCE OF SLOPE

As we strolled along the dirt road, we started down the gentle slope that led toward the eastern shore of Seneca Lake.

Pat was the first to venture a question. "The vineyards on this side of the lake have west-facing slopes. I have read that in northern climates, like those in Germany from which you come, the best vineyards face south. Is there any reason other than the north/south elongation of the Finger Lakes to explain the predominant location of vineyards on east- and west-facing slopes?"

"What you mention about German vineyards is generally correct, but in New York we are considerably further south. In terms of latitude, we are on a parallel with Bordeaux, France. Thus, not only is the sun stronger here, but the more frequent cloud cover also tends to neutralize the advantages of south-facing slopes. In fact, a southern exposure can occasionally be a disadvantage. For example, solar radiation is trapped maximally during the winter. This means that southern slopes tend to lose their snow cover sooner. This has the disadvantage that the insulating influence of snow disappears earlier and the vines prematurely lose their cold hardiness. They are correspondingly more susceptible to the late frosts that frequently affect this region, due to our continental climate. In contrast, east- and west-facing slopes retain their snow cover longer, and provide more protection. Western slopes may even have the advantage of collect-ing additional snow cover during the winter. Southern and eastern slopes may experience the winter equivalent of the rain-shadow effect. The relative effects of these microclimatic influences vary in intensity from year to year and with the wind direction.

"As the ruts in this road signify, slopes also promote water runoff. Although slopes may suffer erosion, the vineyard gains from im-proved drainage. Thus, the soil warms earlier and root growth starts sooner in the spring."

"This is fascinating! Why don't we read about this in wine maga-zines?" muttered Pat, swatting at a pesky yellow jacket.

"I don't know," responded Klaus. "It simply may be that few wine writers have sufficient scientific background. If consumers don't demand more explanation, publishers won't insist on it from their writers.

"On a related matter," exclaimed Dave, "I'm irritated with our magazines for giving inordinate attention to European wines. The United States now makes some of the best wines in the world. It's high time our publications gave our wines more exposure!"

INFLUENCE OF SOIL TEXTURE AND PRUNING

"You were talking about soil a few moments ago," said Pat. "There seems to be a lot of clay here. Aren't the best vineyards stony?"

"Stony soils are important in regions like Châteauneuf-du-Pape and Bernkastel" admitted Klaus. "Because stony soils heat and cool rapidly, they warm surface roots during the day and radiate heat to the vines at night. This is especially important in the fall when the grapes are ripening. However, stony soils are difficult to cultivate and can dry out rapidly. In contrast, loamy clay soils have better water and nutrient retention properties. The crucial aspect is what the major limiting factors of the site are. They can be early frosts, water shortage, or nutrient deficiency, etc."

"This sounds fine, but don't vines need to be stressed to produce superb wines!" asserted Claude.

"Not exactly," responded Klaus. "The relationship between yield and quality is one of the most misunderstood aspects of grape grow-ing. Severe pruning under our conditions only promotes excessive

shoot growth, incompletely ripened fruit, and poor flowering in subsequent years. With our training systems, vine growth is less dense,
leaf photosynthesis improved, fruit bud induction enhanced, and
more energy directed into fruit ripening. Thus, both yield and quality
are increased."

"This sounds too good to be true," reflected Claude. "Isn't this
only propaganda to counteract the European mystique of *terroir*?"

"This is no propaganda!" retorted Klaus. "If there is any, it
comes from Europe by implying that outdated techniques are required to make fine wine."

"However, if planting on fertile soil is beneficial," I remarked,
"why didn't Europeans plant their vineyards on such sites, rather
than on poor stony soils and hillsides we always hear about?"

"Instead of monopolizing the conversation, I'll let Mel respond
to that question. He's more knowledgeable about the historical aspects of viticulture than I."

"Thanks for the compliment," said Mel, "but I'm not sure it's
warranted. It's true that I do have a passion for history. The course I
took with Dr. Loubère at Syracuse on wine history got me hooked
on viticultural history. It was his course that eventually provoked
me to want a vineyard; and here I am, grape grower *extraordin-
aire*." He laughed, arms extended to symbolically envelop the surrounding slopes.

"Many of the coastal and valley regions where European civili-
zation had its origins were fertile loamy soils, easy to cultivate.
They were used, as one can readily understand, for annual food
crops. Cultivation of grapes was relegated to slopes or poorer sites,
where little else of agricultural value would grow. Alternately, vines
were trained up trees, usually at the edges of fields. With the fall of
the Roman Empire, grape production became primarily the preserve
of the Church and nobility. When economic conditions started to
revive, regions close to large population centers had an incentive to
improve their wine quality. This resulted in the development of
grape-growing practices designed to restrict vine vigor and direct a
larger portion of the vine's energy into ripening the crop.

"When grapevines were carried to the New World, they were
often planted in fertile valley soils. Although fruit quality probably
suffered, there was no local market for high-quality wine. Thus,

fruit yield tended to be more important than its winemaking qualities. With modern improvements in pest control, fertilization, and
irrigation, vines became overly vigorous. The old training systems,
designed for dryland farming on relatively infertile soils, enhanced
vegetative vigor rather than limiting it. It took time to realize the
source of the problem and devise new training systems. These systems direct the additional energy of vines grown under optimal
conditions to fully ripen a higher fruit yield."

"May I add a further note to Mel's comments?" said Klaus.
"Although many vineyard regions in Europe were initially relegated
to poorer or drier sites, centuries of fertilization have improved the
soil's nutrient status. For example, Dr. Seguin has found a direct
correlation between the ranking of Bordeaux vineyards and the soil's
nutrient content. Because the soil's improved makeup reflects its
organic content, part of the site's quality results from human activity–adding manure. It is a case of simple economics, if there is such a
thing. When a site becomes recognized as producing better wines, its
wines begin to fetch higher prices. This, in turn, provides the funds to
use practices that can further improve grape and wine quality. As
someone more succinct than I put it: 'Success breeds success.'"

By this time, we had reached the end of the vineyard and were
looking out across Seneca Lake. The maples, poplars, and elms
along the shore were already responding to the shorter days and
cooler nights. Their leaves showed distinct gold and reddish hues. A
light breeze induced the serene sapphire waters to lap gently against
the pebbles on the shore. It was an ideal spot and an appropriate
time for lunch. Some of us had brought a hearty meal with sufficient
and appropriate libation to fortify us for the grape harvest ahead.

As we arranged ourselves on the nodding grass, one could not
help but notice the sparkle from the aluminum strips dancing in the
breeze from adjacent vine supports.

"What are the tinselly strips for, Mel?" inquired Pat.

"They are intended to startle birds and keep them away from the
ripening fruit."

"Does it work?" Pat continued.

"I'm not really sure. We have tried several techniques over the
years, but with limited success. We are planning to try a spray next
year–methyl anthranilate. It apparently is distasteful to birds. Sur-

prisingly, it is used commercially as an artificial grape flavorant in jams, juices, etc. Methyl anthranilate is also a natural constituent in several grape varieties. If it lives up to its promise, we can protect our investment and continue to be environmentally friendly."

INFLUENCE OF ADJACENT WATER BODIES

Klaus remarked, "The breeze that is creating the jostling of the tinsel illustrates another influence of site selection and slope. As the sun warms the land, air starts to rise and cooler air moves in from the lake. When the warm air climbs, it begins to cool and flow back to the lake, replacing the air flowing toward the land. This produces the mild draft we are currently experiencing. In the spring, cold air during the night tends to slide down the slope and out over the lake. As it picks up heat from the water, the air rises. Again a breeze is created. It helps to limit the accumulation of cold air around the vines and minimizes frost damage."

While most of us were listening to Mel and Klaus, Pat had become transfixed on the lone sloop out on the lake. Pat was as crazy about sailing as he was about motorcycles. However, due to financial considerations, his interest had to be limited to a small skiff he kept at the Finger Lakes Marina outside Asbury. The sloop appeared almost to float transcendentally over the lazy water. As the yachtsman tacked the boat, it leaned into the wind and began to create waves that reflected diamonds of light off the water. When Pat came out of his trance and noticed us staring at the shimmering tinsel, he posed a question that had lain hidden for years. "Is there any truth to the story that significant amounts of solar energy are reflected from bodies of water, like our lake here, up into surrounding vineyards?"

"Funny you should ask that," said Klaus, picking a thistle off his pant leg. "I did a paper on vineyard microclimate last spring in Advanced Climatology. As part of the literature search, I read Geiger's classic *Climate Near the Ground*. In it, Geiger refers to several German and Japanese articles on the topic. At the low sun angles in the spring and fall, water can reflect up to 30 percent of the light received by vines. The effect is greatest on sloped sites and in the early morning and late afternoon. However, the claim that heat is reflected is incorrect. Solar heat is too effectively absorbed by water to be reflected off its surface."

"So that means that when I am basking on the skiff, I am not being warmed by the light reflected by the surrounding water," joked Pat.

"Not in any humanly perceptible manner anyway," responded Klaus, removing the last thistle from his trousers.

For some of us, all this talk about the warming action of the sun brought on a feeling of somnolence, or so it seemed, as we reclined on the toasty vegetation. Others, like myself, went off in search of savory mushrooms or to collect butterflies. Dave was an amateur entomologist as well as a wine lover.

After a short siesta, it was time to head back up the hill to the pick-your-own section.

TIMING OF HARVEST

"We are picking today because we have the time and the fruit looks ripe," I said. "Regardless, winemakers must have more sophisticated ways of deciding when to harvest."

"Yes and no," came Klaus's pensive response. "In the old days, it was essentially as simple as your description, with the addition of tasting the grapes. With the discovery of the importance of sugar to fermentation, and a simple means of measuring grape sugar content, this property became the first numerical measure of fruit ripeness."

"Have you ever tasted wine grapes off the vine?" said Mel.

Never having resided in a wine-producing region before, my answer was, "No."

Pat sprang to the task. Reaching out for a bunch from the closest vine, he popped a few grapes into his mouth before passing the cluster around.

"Wow!" exclaimed Pat in his usual exuberant fashion. "These berries are juice bombs. The yeast in the bloom must be chopping their lips in anticipation of attacking the sugars inside."

Klaus laughed. "Quite possibly, but the matt-like covering on the fruit people call the 'bloom' does not consist of yeasts. It is actually a layer of microscopic waxy plates covering the skin of the berry.

"However, to continue the topic of harvest timing, another critical chemical property is berry acidity. Sugar provides the energy for the fermentation that produces the alcohol, but acidity gives the wine its fresh taste. Too much acidity, though, and the wine tastes hard, while too little acidity makes the wine taste flat and characterless. However, unlike sugar content, which increases during maturity, acidity tends to decline. Of these two changes, only sugar content can be quickly and inexpensively measured in the field (with a refractometer); acidity must be assessed back at the winery."

At this point, Mel pulled his trusty refractometer out of his jacket pocket, squeezed a few drops of juice from a berry into the gadget and held the other end up to his eye. Satisfied, he gave it to us to try. "Hold it up toward the sun." Mel illustrated with an imaginary refractometer in his hand. "Adjust the knob on the top to superimpose the two long lines you can see and you will be able to read off the approximate sugar content of the juice."

The refractometer reminded me of a high-tech kaleidoscope. While we were squinting through the device, Klaus continued his discussion of harvesting.

"Depending on the prevailing climatic conditions, sugar content (in cool climates), acidity level (in hot climates), or sugar/acid balance (in temperate climates), are the primary determinants of harvest date. Additional factors include the health of the grapes, likelihood of rain or frost, and labor availability if manually harvested. To collect the fruit while cool in hot regions, harvesting by machine may be conducted either during the early morning hours or at night. This seldom is a concern here. Unlike today, our harvesters often need to wear heavy sweaters and toques when picking, even in the afternoon."

Refractometer

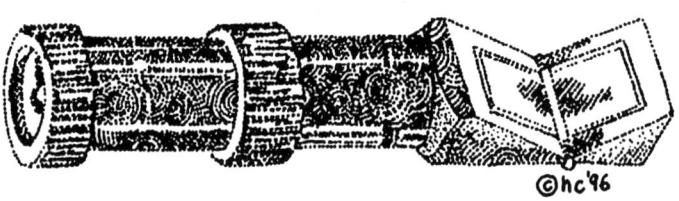

©hc'96

"But what about varietal flavor in the grapes? Isn't that at least as important?" asked Dave.

"It certainly is, but the problem is how to measure it. With some varieties, such as Muscat cultivars, we know that important aroma compounds accumulate in the grape at roughly the same time as does the sugar. There is also a correlation between sugar accumulation in berries and the production of anthocyanins, the pigments in red grapes. However, even in these cases, the correlation is not perfect. With most other grape cultivars, the chemical nature of their varietal aroma is unknown. With additional research and simplification of assessment, it may be possible to time the harvest to achieve a predetermined level of aroma in the wine."

HARVEST METHOD

"This is fine, but what's the difference between mechanical and manual harvesting?" Pat interjected. "Is the talk about great vineyards only harvesting manually just hype to justify exorbitant prices? If I want good mail service I'm not going to consider using carrier pigeons because it was once traditional! Does mechanical harvesting *really* influence wine quality?"

"I wish I could give you a straight and simple answer, but it's impossible. In the first case, it depends on the grape variety. The fruit of some varieties separate well under the shaking action of a mechanical harvester, while others do not. One can lose a lot of juice from broken grapes. Even a small proportion of broken red grapes make it difficult to produce a white wine from the juice, if that were the intent. Sufficient red pigment will be released into the juice from the broken fruit to give the wine a blush color. However, where the grapes separate well, mechanical harvesting provides the option of picking at night in hot environments. Thus, the grapes can be collected while cool, and one can avoid the expense of refrigerating the juice before starting fermentation. Grapes can also be harvested around-the-clock to avoid rainy weather that could spoil the crop. Barring mechanical breakdown, harvesting by machine is also more reliable than seasonal or migrant labor. In practice, the choice between manual and mechanical harvesting often comes down to which is cheaper in the long run. Only vineyards associated with high-priced wines have the option of choosing manual harvesting when both result in wines of equivalent quality. But even here, one has the problem of defining what is 'equivalent' quality. From what I have seen in the scientific literature, grapes harvested by both techniques can produce excellent wines. If there are differences, most wine consumers, and wine judges, can't detect them."

"So it's pretty much 'you pays your money and takes your choice,'" exclaimed Pat.

By the time we'd reached our designated picking area, the place was like a beehive. The exquisite weather had acted like syrup to hordes of would-be winemakers. Although one could rent standard field collecting boxes, roughly resembling laundry baskets, many people had brought their own containers. These varied from milk boxes to wicker picnic baskets to even a wooden back basket, which gave every appearance of being an antique *hotte* from the Mosel.

Everyone was in a jovial mood, befitting the day and an impromptu mini harvest festival. Octogenarians who had shed thirty years were out picking with their great-grandchildren. Their combined activity was mutually beneficial. The clusters beyond the reach of the younger generation matched the lowest extension of the senior partner. The only aspects missing were the songs and obligatory game of catch between grape throwers and oral receivers typically shown in wine travelogues.

By the end of the afternoon, those of us new to grape picking had developed an increased respect for those who do this for a livelihood. We also became cognizant of muscles of which we were hitherto unaware.

Our group had originally intended to have a quick repast at the Pancake House on the campus, back at the university. However, our aching muscles and burned faces informed the older members of the group that a more relaxed evening was required. Therefore, we hefted our vinous treasures onto Tony's trusty half-ton and called it a day. We scheduled the grape crush for the following afternoon.

CARBONIC MACERATION WINEMAKING

Tony's garage had none of the elegance of a Château Margaux, but it had all the paraphernalia for home winemaking. Most of us were eager to become initiates in the honorable ranks of the Ithaca *Confrérie de la Barrique*. Tony came by winemaking naturally, as his family had made wine forever.

A venerable wine press was centrally located in a place of honor in the makeshift winery. Off to one side was an elongated planter, devoid of the flower pots it had once contained. It had been doused with metabisulfite and thoroughly rinsed, but still smelled like a burnt match. The planter was to serve as our crusher. On the bench along the east wall, in a space vacated of kids' toys, were hydrometers, cylinders, and an assortment of various bottles of bentonite, potassium metabisulfite, and oak chips. We were ready to do battle with yesterday's harvest.

Looking at Tony, Claude asked, "What's our first task, chef?"

"Do you want to bring down the 'Maréchal Foch' from the patio," responded Tony. "I put them there to heat in the sun. The grapes need to be warm to undergo carbonic maceration. This is the process used in Beaujolais and several other European regions for making early drinking wines. With more standard winemaking procedures, it takes one to several years for the wine to become drinkable, at least to all except to their creators." He chuckled. "Carbonic maceration is as simple as winemaking gets, except by the kit. It will make our Maréchal 'Beaujolais' Nouveau."

"Don't you need 'Gamay' grapes to make a Beaujolais?" asked Dave.

"Technically yes," said Tony, "but the fragrance that distinguishes a Beaujolais does not come from the Gamay grape. The wine's fragrance comes from the process by which it is made."

"Hold on!" said Pat. "I've always read that it's 'Gamay' that gives the wine its distinctive character."

"Although most wine writers ape this view, French researchers know better," responded Tony. "The uncle who taught me the technique showed me the comments of specialists in a book on carbonic maceration. They specifically stated that it's the process that gives Beaujolais wines their distinctive fragrance."

"Then what's the importance of the 'Gamay' grape?" asked Dave, looking perplexed.

"Give 'Gamay' its due," I responded. "It gives the wine its bitter flavor and mouth-feel."

"Would you like to try a bottle of last year's version?" said Tony. There was a joyous uproar at the offer of wine to wet our palates.

A few moments later we were being served six glasses of wine. Each of us took a glass and began to taste the wine. Dave was the first to comment.

"Well, I'll be jiggered!" he said. "You're right. It does smell like a Beaujolais. I'd never have thought it possible from what I've read."

We were impressed by Tony's wine. It definitely showed the typical raspberry/kirsch character of a Beaujolais.

"This is a lot better than most of the homemade wines I've tried over the years," remarked Dave. "It smells like a Beaujolais, but tastes different. I like the difference, though. It has more body and flavor. Too often I find Beaujolais wines a bit thin in the mouth. Though to be fair, some Beaujolais wines are flavorful, but they certainly cost more than this wine."

Tony said nothing, but looked pleased as punch. After most people had finished their wine, Tony spoke up, waving his hands like a sergeant major. "OK! Enough dawdling! We'll never make any wine if we just stand around here drinking it."

Phil and Dave disappeared and almost immediately reappeared with the 'Maréchal Foch.'

Everyone was all ears, intent on learning about a little-known winemaking technique.

Tony continued, "You add, with a little compaction, enough grape clusters to fill whatever type of open-topped fermentor you have. I use 25-gallon food-grade plastic containers I get free from restaurants. They're particularly useful as they have tight-fitting tops. This is needed to trap the carbon dioxide released by the grapes during fermentation. In the absence of air, the grapes soon begin to self-ferment and produce the wine's characteristic aroma."

"I've heard of people using plastic garbage pails as fermentors," commented Claude. "Isn't that risky?"

"It certainly can be," said Tony. "Toxins can seep into the wine from some plastic containers. You need to be sure that you are using food-grade containers. I could have used oak barrels, but that would have required taking off the top to put in the whole clusters. However, how would I get the top back on? I'm no cooper. Glass carboys would do, but the opening is too narrow."

Diagram for Wine Production

Diagram of Wine Production

WHITE WINE	RED WINE	RED WINE
'Riesling'	'de Chaunac'	'Maréchal Foch'

HARVEST   HARVEST   HARVEST

FIRST WEEK

	TRADITIONAL	CARBONIC MACERATION
		Grape Berry Fermentation
Stemming	Stemming	
Crushing	Crushing	Crushing
Pressing		
Yeast Fermentation	Yeast Fermentation	Yeast Fermentation

SECOND WEEK

	Pressing	Pressing
	Completion of Fermentation	Completion of Fermentation

THIRD WEEK

Racking	Racking	Racking

Months later

BOTTLING   BOTTLING   BOTTLING

WHITE WINE	RED WINE	RED WINE
'Riesling'	'de Chaunac'	'Maréchal Foch'

"I hate to appear like a neophyte, but what is a carboy?" asked Pat.

Tony laughed as he pointed to a group of large glass jugs in the corner of the garage. "Those are carboys," he said. "The neck opening is hardly big enough for inserting berries, let alone whole grape clusters."

After filling several containers, we had the fun of hoisting them onto a wheelbarrow and gingerly rolling it through the door, down the hallway, and into the kitchen. As we entered, we received a resigned look from Tony's wife. She must have been used to this autumnal intrusion of her domain.

YEAST INOCULATION

When we got back, Tony explained that it was necessary to put the fermentor in a warm spot. Although not ideal,[1] the kitchen was the warmest room in the house. "If you can insert an airlock in the top of the fermentor, you can tell that the grape fermentation period is complete when gas stops coming off. Otherwise, you simply leave the grapes sealed for seven to ten days.

"During the maceration period, the grape skins weaken and many berries break open. Wild yeasts occurring on the skins inoculate and begin to ferment the released juice. By the time you open the fermentors next week, the Beaujolais fragrance will be very much in evidence. By then, the grapes will have weakened so much that they will not need to be crushed before being put into the press.

TRADITIONAL WHITE WINEMAKING

"Our next task is to work on our white grapes. We were especially lucky to be able to get some 'Riesling' last week. Now, which one of you fine gentlemen wants to volunteer to do the grape stomp? Regrettably, I have no live musical accompaniment."

There was a period of profound silence as people's eyes either flicked around waiting for someone to speak, or pondered the state of the world on the cement floor of the garage.

---

[1] about 90°F (32°C)

"What? No volunteers?" said Tony, obviously enjoying the situation. "If not, a drawing of straws will break the silence."

So it was fate that chose Pat for the task. Off came the boots and socks and onto the "ceremonial" wash and rinse with metabisulfite. To the music of beating carboys, Pat strode reluctantly into the trough, like a prisoner walking the plank. Our good humor at Pat's plight was tempered by the realization that one of us would be next to slip and slide in the mix of juice, skins, seeds, and stems. Fortunately for Pat, the grapes were fully ripe and the crushing was easy. Also, we were soon fully occupied ladling the crush into the press. Removal of the released juice by ladling made it easier for Pat to crush the remaining grapes. Otherwise, unbroken grapes would have floated to the top and avoided being crushed.

Phil asked, "When should we add the sulfur dioxide?"

"I'm not planning to add any at this stage because the grapes are healthy," responded Tony.

"Isn't that running the risk of oxidation?" questioned Phil.

"People used to think so, but current thought is to permit limited oxidation during crushing and pressing. It favors both the early oxidation and precipitation of tannins that can induce browning in the bottle. If you look at the juice running out of the press, you already see brown oxidation products."

"But, don't you also need sulfur dioxide to prevent the fermentation from being taken over by spoilage yeasts or vinegar bacteria?" questioned Dave. "I've heard they produce some pretty atrocious odors."

"It's possible, but I've prepared a culture of yeasts with which I'll inoculate the juice. It should dominate the fermentation regardless of what strains of yeast may be present on the grapes."

"How do you know that the added strain will dominate the fermentation? Aren't all wine yeasts just forms of the same organism?" commented Pat.

"Yes, most yeast strains that survive and complete fermentation are variants of the wine yeast. However, my uncle, who is a microbiologist, tells me that researchers can precisely identify yeast strains using genetic engineering. With these techniques, it's now possible to follow the types of yeast strains throughout fermentation. In nearly all cases, the added strain dominates the fermentation.

"Roy is really into wine!"

"To make sure that my selected strain would be active and ready to initiate fermentation, I started the culture last night. I added a packet of dry yeast to equal quantities of grape juice and lukewarm water. Rapid activation was achieved by immersing the culture vessel in a large bowl of warm water. Once it became obvious that the yeasts were active (from the developing froth), I added more juice. Once the mixture was clearly active, I combined it with additional juice to fill the largest measuring cup I could find in my wife's cupboard. Sophia accommodates my taking over part of the house at this time of year. My activities save us a lot of money that otherwise would be spent on commercially made wine. We have wine with every meal. With store-bought wine that would be pricey."

"How does the press work?" said Claude. "It seems rather antique."

"It's an antique alright, and a venerable one at that. It was once used in the Horticulture Department at the university to make experimental wine. I picked it up at a bargain-basement price when the university got a larger press.

"It's simple to operate. Once the press is nearly full and the flow of free-run juice slows to a trickle, you add two semicircular wood plates, called the press head. After adding a few wood blocks, you start ratcheting the winch down on the blocks. This forces juice out between the slats of the press cage. When doing this, stand clear of the press. That is, unless you want to be squirted. Uncrushed berries near the slits in the cage can suddenly burst and shoot out like water from Old Faithful at Yellowstone. As the juice flow declines, you periodically need to add extra blocks, and the winch handle soon collides with the top of the press cage. When it becomes too hard to extract additional juice, you unscrew the winch and take the press apart. After crumbling compacted grape remains, you conduct a second pressing. I seldom do a third pressing. The amount of extra juice obtained simply isn't worth the effort."

"How close to the top do we fill the carboys with juice?" called Phil from the corner of the garage. He and I had begun pouring the free-run juice through muslin into a carboy.

"Only about 60 percent full!" Tony yelled. "I want to save room for some press-run juice in each carboy. In addition, we need room for the volume of the fermenting juice to expand. Because the juice contains lots of suspended solids, considerable froth will form during fermentation. I don't want the froth to back up into the airlock and block the escape of carbon dioxide released during the fermentation. When the carboys are about 85 percent full, add about two cups of yeast culture. Fit the airlock and carry the jugs to the basement. The cool temperature there will slow fermentation and help retain a fruity bouquet in the wine."

Once we had nearly completed pressing and mixing the last press fractions from the white grapes, Phil asked Tony what our next step was.

"Once the final portion of juice is inoculated with the yeast culture and the carboys have been placed in the basement, we need to clean the crushing trough."

Basket Wine Press

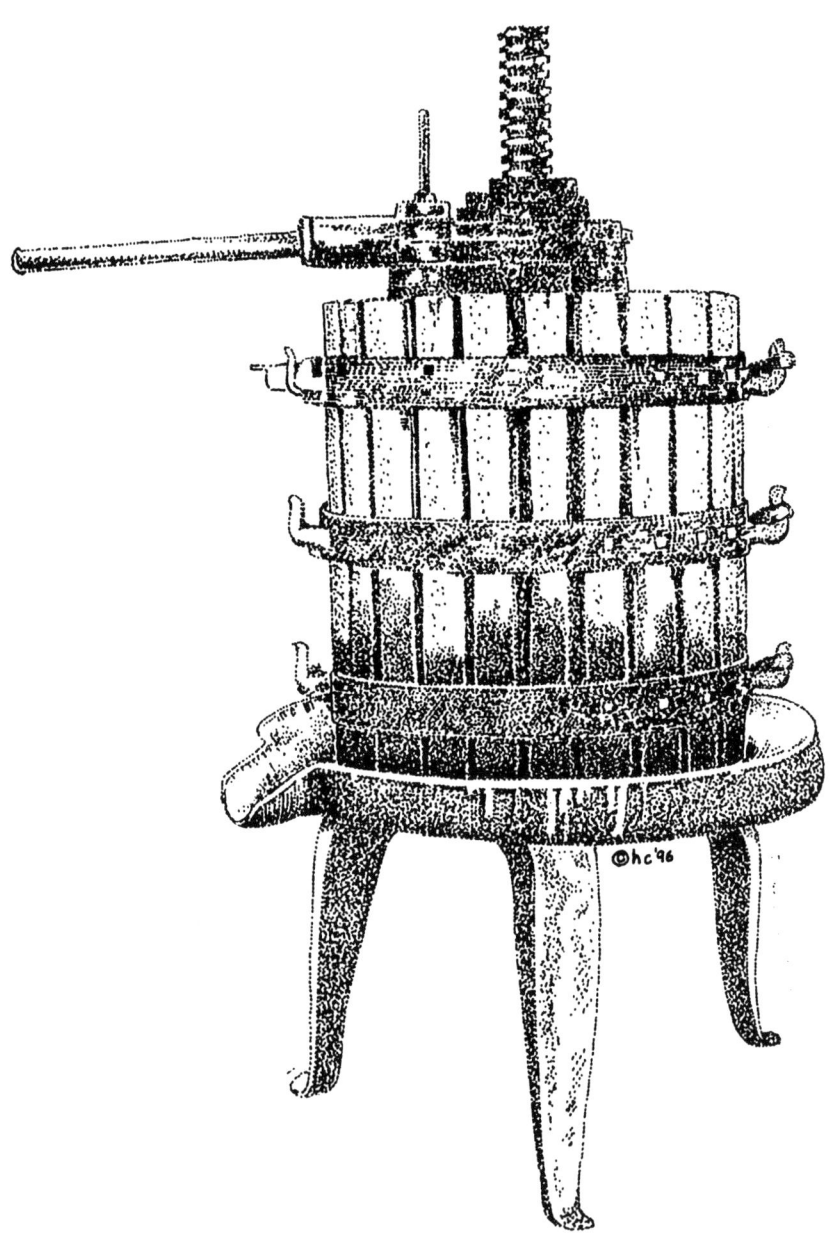

At this, Dave commented, "I'm glad my wife isn't here to see us crushing the grapes with our feet. She'd never drink wine again!"

Tony chuckled. "It is perfectly hygienic. At least as much as when we debone chicken breasts with our hands. Anyway, the acidic conditions and alcohol essentially disinfect the juice. However, the tender-hearted can buy a stemmer-crusher. It does the same job. Regrettably, those available to home winemakers may contaminate the juice with iron. It's not a health hazard, but extracted iron can promote wine oxidation and cloudiness."

TRADITIONAL RED WINEMAKING

As Tony looked around at everyone, save Pat, he slowly enunciated the question we all dreaded. "Do we need to draw straws again for our next grape crusher?"

Phil bravely volunteered to stomp the red 'de Chaunac' grapes, before we had to resort to fate again. His acceptance was greeted with a round of applause. I was wondering how I would have explained red-stained feet and ankles had I been chosen. Suzanne's feeling on the subject of feet-trodden grapes was clear in my mind.

"With the red grapes crushed and inoculated with yeast, there remains little else to do," Tony said. "The 25-gallon-pail fermentors need to be filled only about three-quarters full. The rest of the space is required for juice expansion during fermentation."

"How come we have to leave more space for red than white wine?" inquired Dave, as Phil was preparing himself for the grape stomp.

"Once the crush really starts to 'boil' (ferment), grape skins act like miniature pockets trapping carbon dioxide released by the yeasts. The skins float to the surface and form a cap over the juice. It is this buoyant cap of skin, seeds and yeast froth that increases the volume of the mash. If fermentation is robust, the volume of mash can easily increase 25 to 30 percent. This is less of a problem with the Riesling as we ferment only the juice; the seeds and skins remain in the press."

"Is this the reason why we are not using carboys for fermenting the red wine?" remarked Pat, still cleaning off the last of the sticky juice from his legs.

"There are several reasons for not using carboys at this point," remarked Tony. "The primary one is the nuisance of trying to put the mash through the small opening of the carboy. Even worse would be trying to clean out the seeds and skins after fermentation. In addition, there is no need to seal the fermenting must from air contact as with the white juice. As the cap of seeds and skins forms and carbon dioxide is released, the fermenting must is isolated from exposure to the air. Furthermore, the large opening of the pail makes it easy to punch the cap down periodically. For convenience, I punch the cap down before going to work in the morning, when I get back after school, and again just before turning in for the night. How often you punch down the cap really depends on the person; there is no fixed rule."

"What's the purpose of what you call punching down?" was Pat's next question.

"Red wines are fermented on the skins so that the juice can extract their pigments. In most red grapes, the pigments are located only in the skin. If the skins stay in the cap, pigment extraction is poor. By periodically submerging the cap into the fermenting juice, pigment extraction is improved. In addition, punching down controls spoilage microbes by exposing the skins to the acidity and increasing alcohol content of the fermenting juice."

"Won't there be a problem with fruit flies getting into the cap?" I asked. "Several of those pesky critters have been dive-bombing already. Punching down may drown the beasts, but the thought of fruit fly extract in the wine is less than appealing."

"No problem!" Tony laughed. "I always put cheese cloth over the opening of the fermentor. I have never had to pick flies out of my wine."

Phil was doing a great job. We were going to recommend him for the crusher-of-the-year award, but thought that the less said the better. We didn't want him flinging dripping grape mash at us. Anyway, it was about time to start ladling the mash into the fermentors. The producers of Hellman's mayonnaise and General Foods jams probably never thought their industrial size buckets would live on as red wine fermentors. Once the containers were about three-quarters full, Tony added the yeast culture. He had prepared it the previous evening in 'de Chaunac' juice. We moved the containers to the end of the garage to get them out of harm's way. They were to be left there, even though it would be cool at night. With the garage attached to the house and the door closed, it had proven adequately warm in the past. Also, Tony didn't want to push his luck with Sophia.

About 5:15 P.M. things were cleaned up, more or less, and we broke up to go our various ways. We agreed it had been a fun-filled and informative weekend. We thanked Tony and agreed that I should write Klaus and Mel letters of appreciation for the time they took with us yesterday. We also decided to meet next Saturday to continue our winemaking exercise.

DECANTING

It was amazing how quickly the week went by. Although engrossed in studying the physiology of *Botrytis*, my sabbatical "hobby" was proving a new and increasingly interesting field of endeavor. It seemed that I was always reading or thinking about wine when not involved in my fungal research. It was wonderful to read and think about something other than research. I also realized that once back in Brandon, there would be little opportunity to read freely. In addition, there was such an abundance of wine information at Cornell. How could I not avail myself of the rich opportunity that presented itself?

Due to our faulty alarm clock, I was the last to arrive at Tony's on the subsequent Saturday. My cohorts were already involved in pouring the juice off one of the fermentors. As soon as I got out of the wagon, it was obvious which wine was being decanted. The fragrance seemed to be swirling out of the garage.

"My nose tells me that you must be working on the *nouveau* wine!"

"Correct you are!" Tony asserted. "Would you pass me the graduated cylinder and hydrometer on the shelf? They are to your right. I want to check the sugar content of the juice that is running out of the press."

"What do you expect it to be?" I said.

"It should be fairly low because yeasts have had time to work on it, probably about a Brix reading of 5 or less."

"Brix? What's that?" remarked Claude.

"Brix is a measure of the sugar content of the juice. A Brix reading of 1 corresponds to about to 1 percent sugar."

"What does the hydrometer level read?" I asked.

"It measures about a 5.5 Brix reading. That's OK. It tells me that the process has progressed pretty much as expected. The Brix reading of the juice released upon pressing will be considerably higher because it comes mainly from unbroken grapes. Fermentation of sugars in unbroken grapes apparently stops when the alcohol content reaches about 2.5 percent. Unlike wine yeasts, grape cells are sensitive to alcohol and die at much above 2.5 percent alcohol. Dead grapes don't ferment," joked Tony.

"The juice seems pretty red to me," observed Claude.

"It may look red," countered Tony, "but take the cylinder outside into the light. You'll see that it's hardly more than a dark rosé."

"Hey! How about some help here!" grunted Pat. "These 25-gallon fermentors are no featherweights, you know!"

Tony was clearly right about how the grape skins would become fragile during carbonic maceration. It was far easier to press these grapes than the 'de Chaunac' last week.

"Enzymes released by the dying grape cells break down the glue-like material that holds plant cells together. Thus, the berries burst open even under slight pressure. It's the carbon dioxide trapped inside the fruit that helps to maintain the normal berry shape."

When I looked at the juice coming out of the press, it looked murky.

Dave commented on this feature and wondered if we should use it.

"Absolutely!" said Tony, throwing back his head of black hair. "This is the better juice fraction as it contains most of the flavor. The free-run juice smells nice, but it is the press juice that gives the wine its taste and substance. If you want to tell the difference the press fraction makes, compare a Beaujolais nouveau with a *cru* Beaujolais. The *nouveau* is made primarily from free-run juice and is often thin and watery. In contrast, the *cru* contains more press-run juice and has greater body. This allows the wine to age and improve, something the *nouveau* never does. It is not without reason that Beaujolais nouveau wines should be drunk before New Year's Day."

"Come on, Tony!" I said. "Isn't that just a little bit extreme? I've had *nouveaus* that improved for six to eight months, especially when they were 'closed in' at the beginning."

"Maybe," said Tony, shrugging his shoulders. "You may get better examples up north than we do here in Ithaca."

Getting back to the matter at hand, Phil asked. "Do we have to add a yeast inoculum to the juice as we did with our other two wines?"

"We could, but I never have," responded Tony. "During the week when the grape clusters are sealed in the fermentor, juice escapes from grapes. The juice begins to ferment and establishes its own large yeast inoculum. I let the natural inoculum continue the fermentation."

"So all we do is pour the juice into carboys?" asked Dave.

"Yes, but make sure you mix both the free and press fractions before pouring the juice into the carboys. Otherwise we'll have a few bottles of 'elixir' and an excess of thinish free-run wine. Also, don't forget to fill the carboys only about 75 to 80 percent full. There is a lot of sediment in the juice and the fermentation is usually fast and furious."

While Dave and Pat finished pressing and blending the juice for our 'Maréchal Foch' *nouveau*, the rest of us cleaned the press.

After a final stirring of our traditionally made, red 'de Chaunac' ferment, it was ready to be pressed. It had a gorgeous yeasty-fruity odor. Strong and fragrant in its own right, but considerably different from the *nouveau* with which we'd just been working.

Phil and I began to ladle the still-fermenting must from the plastic fermentors into the press with gallon ice-cream buckets. The presence of the seeds and skins made the ferment too messy to pour directly from the fermentor into the press. That was possible only when the fermentor was about half empty. Much more free-run juice ran out from the 'de Chaunac' than from either the white grapes or *nouveau* wines. Tony explained that this was due to the extensive breakdown of the berries and pectins during the previous week's fermentation.

"I hope you've noticed the darker color of this free run, as compared with that from the carbonic maceration process. This is regardless of the two wines coming from different grapes. Note also the purplish color of the juice. It's characteristic of young red wines."

"It's amazing what you can miss when you're not looking for it," I said.

As the free run of juice slowed, we started to crank the winch down on the press head. The juice initially gushed out, but then slowed, until one cranked some more. Soon it became difficult to pull on the winch lever, and three of us had to hold the press steady while someone tugged on the handle. When it became nearly impossible to coax out any more juice, we reversed the ratchet direction and raised the winch. Then came the job of knocking free the steel pins holding the two halves of the press cage together, followed by the grubby task of breaking the "cake" of seeds and skins. After reassembling the press and adding the seeds and skins, it was back to extracting whatever juice remained.

"Some people take the cake of seeds and skins after the second pressing, add sugar and water, and ferment the material again," said Tony, as we grunted trying to extricate the last niggardly drops of juice. "I don't re-ferment the pomace myself," he noted. "Commercial producers occasionally re-ferment the pomace and distill the product to make a *marc* brandy. It makes a real screech!"

When we had squeezed out the last drops of juice, we mixed the free- and press-run fractions. There wasn't enough to ferment the fractions separately. Tony mentioned that in commercial wineries, the free- and press-run fractions were usually fermented separately. This allows them to control, by blending, the amount of pigment and tannins in the finished wine.

As Phil was pouring the mixed juice into the first carboy he remembered to ask, "Do we only fill the carboys three-quarters full again?"

To Phil's surprise Tony's answer was, "No, you can continue pouring until they're about 95 percent full."

"How come?" responded Pat, looking surprised.

"The difference is that the fermentation is nearly finished, at least in terms of the proportion of sugar fermented. While you fine gentlemen were grunting and groaning with the press, I checked the Brix level of the juice. It's down to a reading of 2 from its original value of 22.8. The final portion of sugar is used very slowly, and may take another week or two to be completely consumed by the yeasts. Right now, the yeasts are quite torpid. They are swimming in a 10.5 percent alcohol solution. It's amazing that they are able to do anything at that alcohol level."

"Is it because the juice volume won't increase much from this point on?" asked Phil.

"Exactly. In fact, the volume may actually decrease," commented Tony.

"Why should it decrease in volume if not much is happening?" asked Dave.

"That's because during fermentation, heat is produced. This, along with the carbon dioxide gas produced during fermentation, increases the volume of the fermenting juice. Once fermentation slows down to a trickle, the heat escapes and the juice cools to that of the surrounding environment. This results in a contraction of the wine's volume. Take the thermometer from the sulfur dioxide solution in the cylinder. It's on the bench under the window. Check the temperature of the juice you are pouring into the carboys. I'm sure you'll find that it's warmer than even the current garage temperature. It was chilly in here earlier this morning."

"It's 5°F (2.8°C) warmer than the temperature of our temporary winery," declared Claude. "Where did you learn all this stuff about winemaking, Tony? I've known other home winemakers, but none of them know what you know. Were your ancestors professional winemakers?"

"No, but I wish they had been! Had they, I might now be a winemaker back in sunny Italy. Nevertheless, I've been exposed to winemaking since I was knee-high to a grasshopper. My father, grandfather, and uncles were always brewing up something. It's from them that I got my basic love of wine and winemaking. The more technical aspects of the subject I got from taking a night course at Cornell. It put my practical knowledge into a theoretical framework, and now allows me to understand what I'm doing. The course also exposed me to different winemaking styles."

"What do we do with our white wine now?" asked Pat, impatient to get things done. He didn't want to lose any more time than necessary in what might be his last chance to get his skiff out on Cayuga Lake this year.

"In fact, there's little left to do," responded Tony. "The white wine is still fermenting in the basement and won't be ready to decant for another week or two."

"What exactly does decanting entail?" asked Claude.

"Like most aspects of winemaking, it's quite simple. Decanting involves removing the wine from the sediment that accumulates at the bottom of the fermentor. It consists of yeast cells, tannins, grape skin remnants, etc. Because the volumes are not great, I siphon the wine from one carboy to another. To avoid picking up the sediment, I use glass tubing possessing a short U-shaped bend at the end. The U-shaped end rests on the bottom of the carboy, keeping the tube's opening above the accumulated sediment. To the other end of the tube, I affix a long section of plastic tubing. By sucking on the latter, I draw wine into the tube. Pinching the tubing prevents the wine from flowing backward as I position it into a second carboy. I put the carboy for decanting up on the bench so that its bottom is higher than the top of the empty one on the floor. This makes sure that siphoning removes the wine without disturbing the sediment."

"Excuse me for interrupting," said Pat, "but if we're through, would you mind if I skedaddle? Gayle is waiting to go sailing."

"Sure," responded Tony. "We can clean up here. If I were your age, had a skiff, and a girl waiting for me, I'd probably not have even shown up."

"Don't do anything I wouldn't do!" I hollered to Pat as he was about to hop on his Suzuki.

Pat looked back and shouted, "Never!"

"Don't some Burgundy producers leave the wine on the sediment for a long time? What they call *sur lies*?" inquired Dave.

"They do, but in their case, the wine is aged in barrels, not carboys. The sediment, technically called lees, is periodically stirred up into the wine. This partially aerates the wine and counteracts the tendency of the lees to produce foul-smelling sulfur odors. However, while limiting the production of sulfur off-odors, it favors the growth of vinegar bacteria. The addition of metabisulfite limits their growth, but I prefer to use sulfur dioxide as little as possible. Anyway, most yeast strains produce some sulfur dioxide during fermentation."

"So this is basically it, except for bottling?" questioned Phil.

"Unless you really want to come for the decanting? From here on there's little I can't easily do by myself. It's hardly worth your making a special trip."

Siphoning of Wine

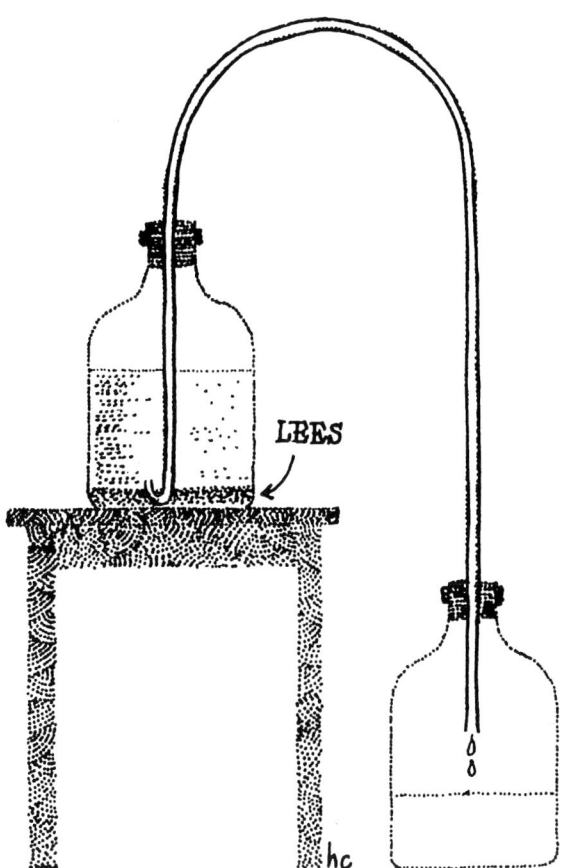

"However, when it's time for bottling and corking, let us know. We want to earn our wine," noted Phil.

"Before we part, do we still meet next at the Heron Hill Winery?" I added.

"What you should have asked was 'Are they still willing to have us?'" Dave chuckled. "This is an especially busy time for them."

"I'll confirm our arrangement to meet with their winemaker and get back to you," replied Phil.

"Sounds good to me," said Dave. "I really like their wines and would love to talk with their winemaker."

"I'll contact him Monday and let you know the exact time."

Chapter 4

Visit to Heron Hill Winery

Heron Hill Insignia. Printed by permission.

Stumbling over mislaid books, I groped my way through the early morning gloom toward the phone in the kitchen.

"Qui pourrait bien appeler à cette heure-ci?" mumbled Suzanne from the bedroom.

"Probably a wrong number," I responded. However, on picking up the receiver it was Pat. After hearing his request, I told him I'd call back in a few minutes.

Upon crawling back into bed, I snuggled up to Suzanne. "Would you mind if we pick up Pat and Gayle on our way to the winery?" I whispered.

"Mais, c'est le *cinq*, n'est pas?" murmured Suzanne, still half asleep. Since getting married, we'd celebrated the fifth of every month—what we call our monthanniversary. Because of the sabbatical, our celebrations had become more low key. Our plan had been to go to the Beaujolais Restaurant in the Triphammer Mall. However, if Pat and Gayle came along with us, our intimate outing could be in jeopardy.

"Something's wrong with his Suzuki," I said. "They've no way of getting to the winery. Pat was wondering if we could pick them up on our way."

"What do you think?" asked Suzanne.

"You'll like Pat!" I asserted. "He's a barrel of fun. Gayle seems very nice too, although I've only met her once."

Suzanne searched my eyes to ferret out what I wanted. Sensing my wish, she replied in the affirmative.

"I know you won't regret it," I said reassuringly.

"I know that," she replied. "Anyway, if we don't celebrate today, we can go tomorrow. The restaurant *is* open for lunch on Sundays?"

"I think so," I said hesitatingly, as I wondered about my response. "If not, we can always go to the Pancake House."

The Pancake House always seemed to be open. If we went there, we could also go for a stroll around Beebe Lake. It possessed a small but captivatingly beautiful nature walk, all within the bounds of the campus.

The foresight of the Women's Alumni in setting aside this tract as a nature preserve in the 1880s has always intrigued me. At that time wilderness was generally viewed as something to be feared, subdued, domesticated, not conserved.

In a flash, I was back on the phone getting directions to Pat's place. By now it was hardly worth going back to bed. That is, if we were to be at the winery by 10:30 A.M. So, we got up and prepared breakfast.

For the occasion, the winery had arranged a lunch on the premises. I knew that they couldn't do this for every group. Thankfully, Phil knew their winemaker, Bob Phillips. Thus, we were being given the royal treatment.

It was shortly after nine when we stopped outside one of the big old mansions on Court Street. Like many of the others in this part of town, it had been subdivided into student apartments or converted into fraternity houses. Pat and Gayle were sitting on the railing of the veranda as we drove up.

As Pat and Gayle climbed in the back seat, Suzanne pointed to the precipitous slope ahead and inquired, "What do you do during the winter? We've heard horror stories about the snowfall here." Coming from Québec, Suzanne was all too familiar with the paralyzing effects of *une grosse tempête de neige*. She wondered how anyone could navigate such streets in winter.

Gayle responded that it wasn't as bad as it looked. At least the cross streets periodically provided horizontal planes. Also, the brick paving gave better traction than asphalt. Finally, the city provided sand boxes along the streets. Nevertheless, we thought it would be prudent to take Highway 35 downtown during the winter.

STEMMER/CRUSHER

Heron Hill is located on the east-facing slope of Keuka Lake, just south of the Vinifera Wine Cellars of Dr. Konstantin Frank. He had been the first vineyard/winery owner to successfully cultivate *vinifera* grapevines in New York. Growing European grape varieties had been tried many times before, but they all had failed due to the combined attack of local grapevine diseases and phylloxera. Grafting and modern fungicides had tipped the balance in favor of *vinifera's* survival.

Although the winery is located on the western side of Highway 54A, most of the vineyards are to the east of the road, where the land slopes gently down to the waterfront. The slopes are much less precipitous than those of the German Mosel, but for a displaced easterner located on the prairies, the rolling hills and shimmering lakes resembled paradise.

Bob Phillips, the winemaker at Heron Hill, was talking with Dave and Phil in the parking lot as we drove up the lane. After introductions, we went into the winery tasting room to await the others. During the interval, Bob brought out a bottle of their most recent Selected-Late-Harvest Riesling. It had an undeveloped but still delicious apricot fragrance, combined with aspects of rose and pine. The lingering flavor on the palate showed its great aging potential. However, it was not possible to follow the wine's development for long, as our other members soon arrived. Although I was ready to continue sampling the wine, everyone else seemed anxious for the tour, so off we went.

Our first stop was the collection bay where freshly harvested grapes were being received. Here, Bob showed us their latest stemmer-crusher.

"It removes the stems (stalks of the grape cluster) and crushes the fruit in one continuous operation," he said. "Depending on the type and style of wine desired, we either press the juice immediately, or allow it to remain with the seeds and skins for several hours. The latter allows flavor, pigments, and tannins to escape from the skins and seeds into the juice."

"You should have seen the crusher we used three weeks ago." Dave laughed and pointed to Pat's and Phil's feet.

Bob chuckled as he replied, "Here we don't use equipment more than ten years old. Your crushers are outdated."

"You mean you don't have wild bacchanalian revelries while crushing the grapes?" joked Tony, joining in the general merriment.

"Feet may be well-padded for the job, but labor costs make their use prohibitive, even if the Department of Health approved!" said Bob.

After watching the machine voraciously devour several loads of 'Riesling' grapes, there was a lull in activity. This gave us a chance to inspect its functional parts. Although not mechanically inclined, I've always been intrigued in the ingenuity of those who design machines. Thus, it was fascinating to get an opportunity to climb up on the apparatus, and stick my head in its mouth and peer down the gullet. The stemmer/crusher had a horizontal rotary shaft on a central axis. The shaft had arms going to three semi-flexible paddles running the length of the apparatus. The paddles rested against a stainless steel cylinder resembling a giant tubular sieve.

When operational, the rotation of the central shaft forces grape clusters against the outer cylinder. Grapes caught in the holes of the cylinder break, releasing their juice, seeds, and skins through the openings. Unbroken berries burst when propelled against the outer encasing cylinder. The stationary outer cylinder collects the juice, seed, and skins, and directs their flow into a catch basin. Because the cylinder holes are too small for the stems to pass through, the paddles drive the stems to the exit portal at the end of the machine.

Internal View of Stemmer-Crusher

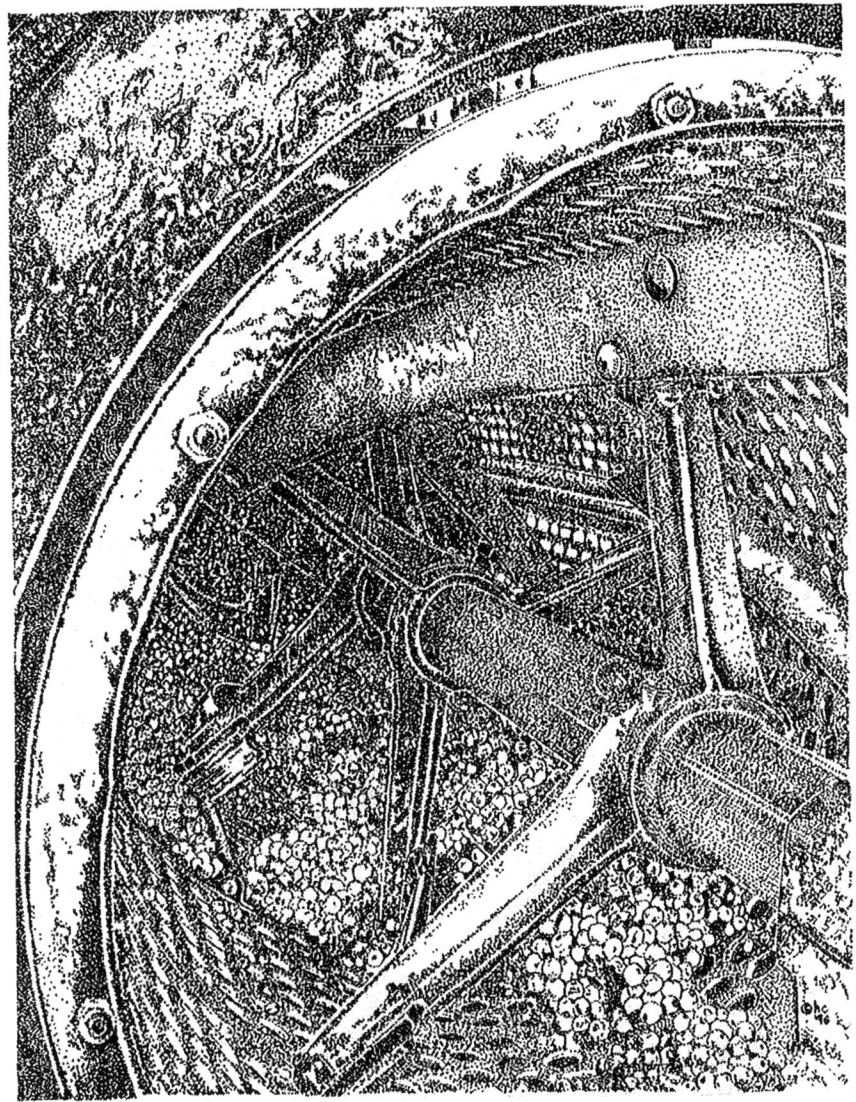

Their other crusher reminded me of my mother's old ringer washing machine. In it, the grapes were crushed between two rotating rollers. The distance between the rollers could be adjusted for berries of different sizes to avoid crushing the seeds.

PRESSES

Because a tractor laden with green-gold 'Riesling' had just arrived from the vineyard, it seemed an auspicious time to move on to the press room. Bob was especially pleased with their Willmes pneumatic press. He explained that it extracted the juice with a minimum of pressure. That meant that the juice was removed with little contamination with grape solids.

"What exactly do you mean by 'grape solids'?" I asked.

"Grape solids refer to the suspended particulate matter found in juice after crushing or pressing," said Bob. "If you wait long enough, they'll settle out, or you can speed the process with centrifugation or filtering. However, I prefer to limit their uptake by using our pneumatic press."

"Because you want to limit their presence, are grape solids undesirable?" asked Dave. "We made no attempt to remove them when we made our wine a few weeks ago."

"Grape solids are desirable in moderate amounts. However, when excessive, they favor the production of off-odors and promote rapid fermentation. To produce the fresh fruity white wine we want, a slow steady fermentation is required. On the other hand, without sufficient soluble solids, the fermentation may stop before all the grape sugars are consumed–a situation called 'sticking.' Restarting the fermentation of a stuck wine is difficult, and the wine is particularly susceptible to bacterial and yeast spoilage."

USE OF SULFUR DIOXIDE

Pat said, "What are your feelings about sulfur dioxide? Tony here prefers to avoid it."

"Well, Tony is up to date. Modern winemakers use it only when necessary. This change reflects recent discoveries that excessive use of sulfur dioxide can promote, rather than reduce, oxidative brown-ing of wine. The change also reflects concern about the possibility of unreasonable restrictions on sulfur dioxide use being imposed by the government."

"So there are times when you would use it?" questioned Pat.

"Definitely!" responded Bob. "It's our best and safest antimicro-bial. It's also useful as an antioxidant. I add a small amount to our finished white wine just after fermentation (on the first racking) and again immediately before bottling. That ounce of prevention affords some protection against the financial losses associated with accidental oxidation. The other time we use sulfur dioxide is if some of the grapes are moldy. For our premium wines, where the grapes are picked manu-ally, we can remove infected grapes. However, with mechanically harvested grapes, this option is not possible. In that situation we add sulfur dioxide during crushing to retard the action of undesirable yeasts and bacteria. . . . Are there any more questions?"

"If sulfur dioxide is so safe, why is the government against it?" asked Claude.

"As you may know, this is a sore point with many winemakers. Admittedly, some producers were overzealous in their use of sulfur dioxide in the past. However, this currently is rare. Although some asthmatics are sensitive to sulfite, this does not justify the negative press sulfur dioxide has received. Other compounds could be used, but they are less effective, more expensive, and/or potentially more harm-ful. Banning sulfur dioxide would undoubtedly lead to the more wide-spread occurrence of poor wine. Moreover, all wines contain sulfites, even if sulfur dioxide is not added. Yeasts produce sulfur dioxide during fermentation, though the amount varies considerably with the strain used."

Bob asked if there were any questions about the Willmes press.

"I have one," I said. "How does this, uh . . . pneumatic press work?"

"Quite simply, in fact. Instead of the press being upright, such as the basket-type press used by most home winemakers, the pneumatic and other modern presses are positioned horizontally. In addition, they can be rotated. That means that the press opening can be posi-tioned on the top for filling, and inverted for easy emptying. It also allows the whole press to be rolled to break up the press cake. You

probably know about the nuisance of disassembling the basket presses between pressings, so you can imagine the advantages of its
avoidance on a commercial scale.

"When the must (juice, skins, and seeds) is poured into the press,
one waits a few minutes for the free-run juice to flow out by gravity.
Notice the openings between the slats that make the outer wall of
the press. Once the flow of free-run juice has dwindled, the trap of
the press is closed and air is pumped into the central air bladder that
runs the length of the press. If you climb the stairs to the platform
on your left, you'll see the air bladder. Inflating the bladder exerts
an even pressure over a large surface area. This allows juice or wine
to be squeezed out under low pressure.

"The one significant disadvantage of the pneumatic press is that it is
not continuous–separate loads of must have to be pressed one after
another. This usually is not a problem with premium quality must, but
adds considerably to the labor cost of producing standard quality
wines."

"How do you press standard quality grapes to keep costs down?"
inquired Pat.

"If your operation is large enough, you'd probably have a 'continuous press.' As the name suggests, the press runs without periodic
interruptions for loading and emptying. The crush goes in one end,
juice is extracted along its length, and dry skins and seeds are ejected
from the rear. With the newer continuous models, it is now possible to
get better control over the quality of the juice than in the past. With a
sequence of drains along the press, the early released juice (better
quality) can be isolated from the later fractions (poorer quality). Subsequent blending can be done, if desired, to the specifications of the
winemaker.

"Any more questions? . . . If not, let's move on to the fermentation
room. Please keep asking questions. They point out aspects I may have
missed."

FERMENTORS

On entering the fermentation room, Pat said, "This place gives
more the impression of a surgical room than a winery! Where are all
the cobwebs, mold, grape stains, . . . the atmosphere?"

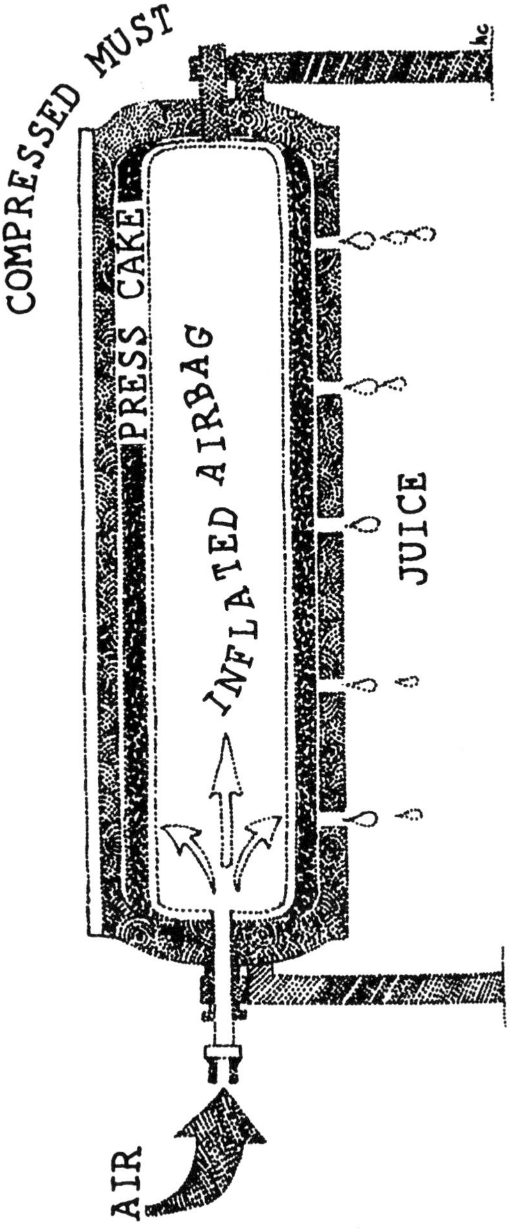

Pneumatic Press

COMPRESSED MUST

PRESS CAKE

INFLATED AIRBAG

JUICE

AIR

"That type of atmosphere went out long ago. We don't want to risk microbial contamination. We'd be out of business mighty fast if we lost thousands of gallons of wine to the likes of *Brettanomyces* or *Acetobacter.*

"What on earth are those?" commented Claude. "Germs from outer space?"

"Even if they were, the wine would be no less saleable. Contamination with *Brettanomyces* is one of a winemaker's worst nightmares. It can quickly give the wine a stench resembling rotten onions. It is all the more a concern because of this yeast's comparative insensitivity to sulfur dioxide, our best and safest disinfectant. *Acetobacter* is not much better. It is one of the acetic acid bacteria. Given much of a chance they turn wine into vinegar, or at lower levels, give the wine a sour ethyl acetate odor."

The size of the fermentors was impressive. I'd seen photos of fermentation tanks, but this was my first encounter with them in-the-flesh, so to speak. Bob allowed us several minutes to walk around, gawk at the equipment, and adjust to the space before discussing fermentation.

"We use three 20,000-gallon, foam-insulated, refrigerated, stainless steel tanks, and several smaller 1,000-gallon tanks. Because we make primarily white wine, and prefer cool fermentation temperatures, all our larger fermentors are insulated and temperature controlled. Surface insulation limits heat exchange to and from the buildings. This maximizes the efficiency of the coolant circulating between the inner and outer jackets of the fermentor."

"Why do you need artificial cooling when this gymnasium of a fermentation room is so cold? I'm glad I have a sweater and kept my coat," commented Gayle.

"Although the room feels damp and cool, the large volume of juice produces considerable heat during fermentation. Heat eventually accumulates more rapidly than it is lost through the sides of the tank. The result would be an unwanted rise in temperature of the fermenting juice. Thus, we need coolant to maintain the desired temperature throughout fermentation."

"I understand that, but where does the heat come from?" Gayle asked.

"When you are doing aerobics, does all the jumping and arm flailing keep you cool? No! During exercise, your rate of metabolism goes up, and so does your heat production. When fermentation cranks up, yeast metabolism also generates a lot of heat. There are about ten to one hundred million yeast cells per milliliter of juice. That's equivalent, in numbers of yeast cells, to the world's population in a teaspoon of juice. Thus, there is no shortage of microscopic heat generators in fermenting juice."

"You mentioned that you keep your white juice cool during fermentation," commented Dave. "We did the same when we made our white wine several weeks ago. At the time Tony explained the reasons, but I've forgotten. Also, why didn't we produce the red wine at similar temperatures?"

Bob looked at Tony, but Tony pointed his outstretched hand toward Bob, saying, "It's your stage. We came here to get your views, not mine."

"I'm sure Tony mentioned the current preference for fermentation temperature between 50° and 60°F (10° to 15°C). This slows fermentation and favors the production and retention of esters that give most young white wines their fruity character.

"Temperature control is equally important with red wines, even though the temperature range is considerably higher, usually between 75° and 80°F (24° to 27°C). If the temperature rises much above 90°F (32°C), the yeast cells become torpid and may die. The higher temperature range preferred for red wines reflects the importance of extracting red pigments and tannins. Higher temperatures also favor the production of glycerol, which may improve the smooth sensation of wine. The more limited production of fruit-smelling esters at the higher temperatures is generally of little significance, because of the greater flavor of most red wines."

"Are the smaller, noninsulated tanks for your red wines?" inquired Claude.

"We don't make red wines," commented Bob. "The use of the small fermentors depends on the volume of must or juice we have. For some of our premium or special late-harvest wines, we don't have enough juice to justify tying up one of our larger fermentors. The space over the wine also increases the likelihood of oxidation, even if we attempt to completely fill the void with an inert gas such as nitrogen. Smaller fermentors also allow us to keep various lots of juice separate for potential blending after fermentation. If the lots were mixed before fermentation, our options for selective blending would be lost."

CHAPTALIZATION

"There's a lot of rancoring about the effects of adding sugar to the juice before fermentation of some European wines," commented Dave. "Some writers imply it's deceitful, while others state that it's warranted. Do winemakers in the States add sugar?"

"With American and French-American hybrid grapes, such as 'Concord,' 'Niagara,' 'Cayuga,' 'Vidal,' etc., it often is necessary. The fruit rarely reaches the sugar content necessary to generate the standard alcohol content of 11 to 12 percent for table wines. In poor years, we may even have to add sugar to our European *vinifera* varieties. However, in most years this is unnecessary. For example, our 'Riesling' grapes possess enough natural sweetness to produce the alcohol content we want (10.5 to 11 percent). Our premium wines are never chaptalized–the term used for the addition of sugar before or during fermentation. Sugar is not added to increase the sweetness of the wine, but to feed the yeasts and increase the wine's alcohol content. The sugar is all consumed during fermentation. Note also that chaptalization cannot correct for the absence of flavor or color in unripe grapes."

"If chaptalization, as you call it, is used simply to increase the alcohol content of the wine, what's all the fuss about?" asked Dave.

"The 'fuss' involves publicity. Warm regions, usually harvesting fully ripened grapes, crow about their banning the process, while cooler regions, often plagued with incompletely ripened fruit, permit the addition of sugar. Furthermore, some countries designate, albeit indirectly, the use of chaptalization on the wine label; others do not note its use at all."

"Why is this?" asked Gayle.

"Beats me?" said Bob with hands upraised. "Nonchaptalized German wines are identified by the terms QmP or Prädikat, while French wine labels remain silent on the matter. Nevertheless, the producers who employ chaptalization must register the fact with their respective government. There is also a maximum amount of sugar that can be added. These regulations were put in place to give the government some clout in convicting unscrupulous producers who 'stretched' their juice with sugared water."

"Where does the term chaptalization come from? It's a mouthful," commented Dave.

"It comes from the name of an early nineteenth century researcher, Jean-Antoine Chaptal. He advocated the addition of sugar to produce better, more stable wines in poor vintages. He was right. Regrettably, as with most technical advances, there are those who misuse the technique."

YEASTS AND FERMENTATION

"Do yeasts do anything other than just produce alcohol and heat during fermentation?" asked Suzanne, in her soft French accent. She had been having a short confab in French with Claude.

"Oh! Gosh yes!" Bob replied. "But a more appropriate question might be 'what *don't* yeasts do during fermentation.'"

After a pause to collect his thoughts, Bob continued. "Except for the sweetness, sourness, bitterness, and astringency of wine–which all come from the grapes–most fragrant compounds in wine are generated by yeast action. A wine's bouquet comes either directly from yeast metabolic by-products, or from their modification during aging. Except for some special grape varieties that possess distinctive aromas, most of what you smell in a wine comes from yeast action. In addition, the alcohol produced during fermentation helps dissolve pigments and tannins out of the skins and seeds. Furthermore, alcohol and other yeast by-products are vital in giving wine much of its aging potential. Grape sugars may be the fuel, but yeasts are the engine."

"Do you use separate yeast strains for fermenting your different wines?" asked Phil.

"Generally no," Bob responded. "We use a commercial strain that we've found works well for us. Because yeast strains are as unique as people, we initially tried many strains. Eventually, we found one that ferments well at low temperatures, produces a nice complement of fruit esters, and is a low sulfite producer. It also has a property that kills most wild yeasts that come in with the grapes.

"Are there any more questions here? . . . If not, I'd like to show you the cellar where we mature and age our wines."

While we were progressing toward the cellar door, Dave remembered that he had wanted to ask about a shiny machine sitting next to another odd-looking piece of equipment. He called out to Bob, who was just about to slide open the door into the cellar. "What's the function of these pieces of equipment here?"

"The centrifuge and plate filter?" asked Bob. "I didn't mention them as I thought no one would be interested. The centrifuge–the large kettlelike machine next to you–can be used at various stages in winemaking. For example, if the level of grape solids in the juice were higher than desired, the excess could be quickly removed with the centrifuge. Alternately, centrifugation can be used to clarify wine for early bottling. The plate filter–the other piece of equipment–also can be used to remove particulate matter. Would you like me to discuss how either of them works?"

Based on the studious silence, it appeared that everyone's interest had been satisfied. "I suspect that most of you are starting to become more interested in the forthcoming tasting and lunch than the finer points of wine clarification." Bob chuckled. "Let's move on to the cellar, our last stop before lunch."

MALOLACTIC FERMENTATION

The cellar contained a large collection of stainless steel tanks of various sizes and a selection of oak casks, from 50-gallon barrels to 1,000-gallon ovals. This room smelled the way we thought a winery should smell. The fragrance of wine and wood aromatics escaping from the oak casks was heady. It seemed to put everyone in a jovial mood.

"I'm sure this is your favorite part of the winery," declared Bob. "This is where modern equipment is less evident, oak cooperage abounds, and you are nestled by the bouquet of aging wine. In other words, it has the 'atmosphere' that Pat found missing earlier. All we need are mold-covered stone walls, cobwebs, and flickering candlelight to transport us to some old cave in Europe. It is here that the artistic aspect of winemaking reaches its zenith. Science tells us much about what is happening during maturation and aging, but it's still on the palate of the winemaker that the final character of the wine is made. Although much of our wine is matured in stainless steel to retain its inherent character, our premium wines benefit from some exposure to oak.

"During maturation, particulate matter settles out, the wine loses its yeasty bouquet, excess carbon dioxide escapes, and the various flavor components harmonize. Various procedures may be employed to promote these changes and improve the character of the wine. These may be as complex as malolactic fermentation, or as simple as adding bentonite to encourage spontaneous clarification. The most well known of these procedures to wine buffs is in-barrel maturation, to add vanilla-like subtleties to the wine.

"Of processes that may occur during maturation, winemakers disagree most about the relative merits of malolactic fermentation. It is a second bacterial fermentation that often follows the initial yeast (alcoholic) fermentation. Malolactic fermentation softens the harsh sourness of acidic wines by selectively converting malic acid into the less sour-tasting lactic acid. However, in wines that are low in acidity, malolactic fermentation can give the wine a flat taste. Even worse, some of the bacteria involved can generate unpleasant odors, notably a smell resembling sauerkraut. Finally, malolactic fermentation is difficult to induce in acidic wines that would benefit from its occurrence. In contrast, malolactic fermentation occurs easily in wines low in acidity that are often spoiled by the process."

"Is this the second fermentation home winemakers talk about?" questioned Claude.

"No. It's quite different. The second fermentation of home wine-makers refers to the slow phase of fermentation, typically occurring after racking. The vigorous 'boiling' phase of yeast fermentation is what they call the primary fermentation."

"What is racking?" inquired Gayle. "Surely it can't have any-thing to do with the rack of medieval dungeons."

"It certainly doesn't!" Bob laughed. "Racking refers to separat-ing the wine from the sediment that accumulates. It consists mainly of dead and dying yeast cells, but may also contain grape skin and cell remnants, as well as grape seeds."

MATURATION IN OAK

"Although a few of you may have heard of malolactic fermentation, I'm certain that most of you know about maturation in oak. This is one of the more intriguing aspect of winemaking. Oak maturation provides us with one of the more dramatic ways of modifying a wine's character. It is like a chef choosing what spices to add to his sauce. The winemaker may use oak to really spice up the wine, or simply to emphasize certain aromatic nuances. Alternately, the winemaker may accentuate the wine's subtle flavors by aging in stainless steel. To me, when critics complain about wines being too oaky, it's like people griping about spicy food. Whether exposure to oak improves the wine depends on its flavor, the views of the producer, and, of course, the preferences of the consumer. Regardless of what some critics say, there are no hard-and-fast rules about taste."

"That's certainly a different presentation of the issue than I've ever heard before," reflected Suzanne. "Although Ron has told me that the oak character of a wine depends on how much the oak had been heated during barrel making. From what you have just said, it sounds like the differences might be comparable to the differences between white and black pepper."

"I'd certainly agree with that," remarked Bob, pleased at the extension of his spice analogy. "The species of oak, rate of wood growth, and manner of drying all impart subtle differences to the flavors present in the wood. Oak adds a whole palate of flavors that can enhance, or mask, the characteristics of a wine. Some winemakers are artisans, but others are just skilled technicians.

"Well, I suspect that all this talk about spices has whet your appetite for some good food and wine. So, if you'll adjourn to the patio, I am sure lunch is ready. Phil, you know how to get there." He turned to the rest of us. "I'll leave you in Phil's hands while I check out the collection in the winery's private cellar."

The patio on the northeastern side of the winery was admirably situated to exploit the panoramic view up and down the valley and the southern branch of Keuka Lake. From this vantage point one could envision the tuning fork shape of the lake.

Frosts had yet to damage the cerise-colored Lady Washington geraniums or the white-eyed blue lobelia. The flower boxes rested on the chocolate-stained railing that bounded the patio. Particularly appealing were the trailing cushion mums cascading down from each corner of the deck.

The clouds that had closed in earlier in the morning had largely dispersed and the patio was bathed in solar warmth. Although the setting was gorgeous, I was a bit embarrassed by the furtive glances we got from regular visitors coming and going from the tasting room. I was glad when Bob arrived with his arms full of wine bottles and we got down to the sensual pleasures of the table.

Initially, I was surprised at the royal blue tablecloths that draped the tables set out before us. However, when we sat down, the wisdom of avoiding white became evident. We would have been blinded by the brilliant solar reflection off white sheets.

I did not know where the lovely lunch materialized from, but our appetizer—canapé of smoked salmon—seemed to appear as if by magic. It was served with their dry Gewürztraminer. The trace of bitterness and litchi flavor united amazingly well with the salmon. Phil was especially pleased with the combination, as he had suggested the pairing. The main course was thinly sliced turkey breast sauteed in butter, its sauce made by deglazing the frying pan with white wine and lemon juice. The light dredging of the breast slices in flour gave the sauce just a touch of thickening. The succulent juices of the poultry seemed to find a perfect marriage with the sauce. The turkey was served on a bed of saffron-flavored Basmati rice with butter lettuce and kumquats. The meal harmonized perfectly with their premium, barrel-fermented, oak-matured Chardonnay. When we savored the culinary exposition before us, we had to applaud the chef, who turned out to be Bob's wife. Later, we found out that she was a recent graduate from the Hotel School on campus.

For dessert, we were served a bottle of their first experimental selected-late-harvest Riesling. It had the smooth opulent richness to replace dessert but had sufficient acidity to prevent it from being cloying. Even the golden sheen seemed to blend with the fall colors that clothed the hills rising from behind the winery.

After many toasts to Bob, his wife, and the winery crew, it was time for Bob to leave. He had paperwork on this year's production to complete for the Bureau of Alcohol, Tobacco, and Firearms. So it was with sincere regret that we bade farewell to our gracious host and hostess. Before going our separate ways, we tarried in the tasting room to purchase some special weekend wine.

Because the afternoon was still young, we wondered if Pat and Gayle wanted to go hiking up the trail along Buttermilk Falls, at the base of Cayuga Lake. They thanked us, but said they would prefer if we would drop them off at the Suzuki dealership near the Commons. It was only five minutes from our destination–Buttermilk State Park–just outside Ithaca.

It had been years since we had had a chance to scuff our feet through piles of leaves, and glory in the forest inflamed in surrealistic crimsons, golds, and yellows. The sensation obtained has always been bittersweet. The joy of walking through a fall calendar is tempered by the realization of what will happen with the first heavy rain. The virgin white plumage of the forest garbed in snow is preceded by the dismal darkness of the arboreal skeletons of late autumn.

Even after our long walk up the groomed path along the seemingly endless series of cavorting falls, we did not feel hungry. Instead, we relaxed on the lawn with an assorted collection of toddlers, romantics, truck drivers, immigrant families, and students–all intent on enjoying the last summery weekend of the fall. As the sun began to set, and the dampness to rise, it was time to head back to the apartment. Tonight's repast was to be Ithaca burgers[1] and Chianti.

---

[1]Lean ground beef flavored with meat spice and combined in equal quantities of bread crumbs. The patties are divided and a slice of processed cheese inserted in the middle. These are pan fried and otherwise treated like hamburgers.

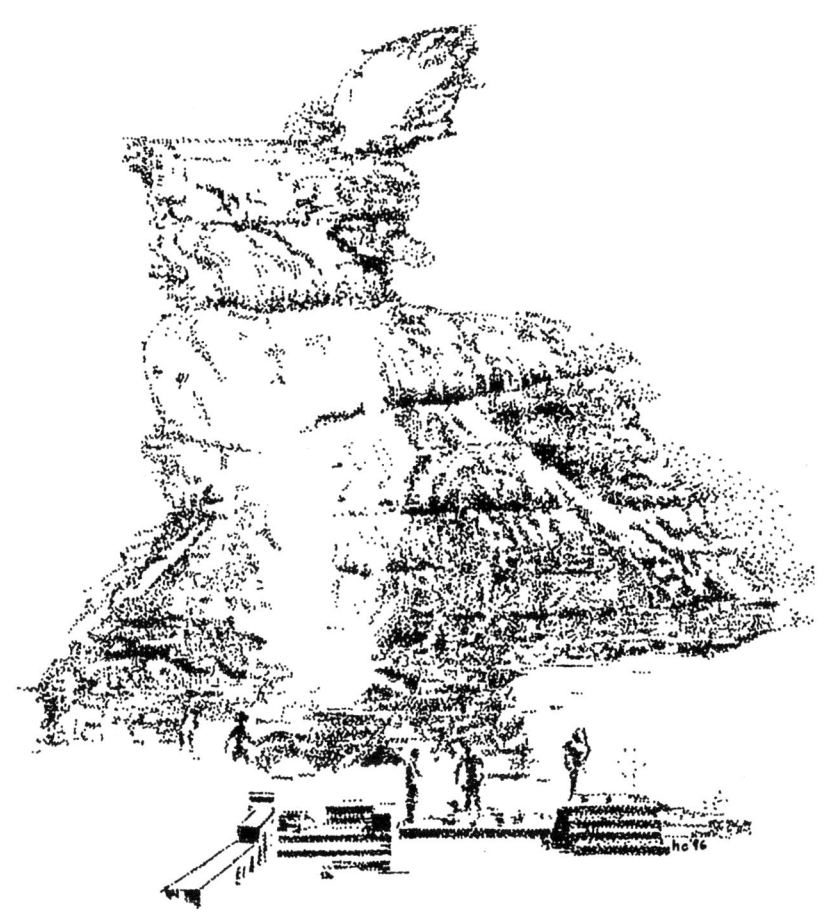

Buttermilk Falls

Chapter 5

Views on Wine Selection

It was a month before I attended our next meeting. I had missed the first racking of our wines because it had fallen on the Canadian Thanksgiving Weekend. We had taken the opportunity to visit my family in Ontario.

This evening's topic was wine selection. It was being given by Peter Maples, the owner of a wine shop just off the Commons. Suzanne and I had met him several weeks earlier, when we had been out scouting the various bottle shops in town.

I was being continually surprised by the diversity of Ithaca's liquor stores. It was such a radical change from the former monotonous similarity of the liquor stores in Canada. I still remember when products were not even out on display. If you wanted a product, you had to look it up on a list, fill out a form noting its name and number, price, quantity desired, your name, age, date of birth, address, telephone number, social insurance number, etc. On presenting the order, you would be informed if it were available, and where you could pay for it. With your paid receipt and stamped order, you could proceed to collect your item. The product subsequently appeared out from some cavernous warehouse. There was no romance about purchasing wine in the old days!

What especially delighted me about Ithaca was the knowledgeable wineshop owners, proud of their selection. Some precious items might be stashed away in the back, available only by asking. The limited stock was reserved for special clientele who got wind of it through the local grapevine. Occasionally, though, prestigious wines might be placed unsurreptitiously on the floor, not more than inches from your feet. I still remember the bottle of Château d'Yquem I almost sent flying at East Hill Liquors one Saturday morning. Thankfully, I was too intent on inspecting each bottle to be moving quickly enough to have caused much damage.

It was at the shop off the Commons where I first met Peter Maples. Peter had emigrated in his early twenties from Surrey, England, just after the Second World War. His accent still registered his southern English origin. I always enjoyed the formality, but genuine warmth, of his gentlemanly greeting. It made me feel appreciated and important, even if I were attired in jeans or shorts.

Peter noticed us eying his *cru* Bordeaux and Burgundy wines. We admitted our interest in these wines, but conceded that they were beyond our purchasing power. At this, he directed our attention to one of a series of barrels where wines from diverse countries were displayed. He suggested that if we liked aged Bordeaux, then we would probably enjoy the wines from Rioja, a region in northern Spain. Although I had never heard of Rioja, nor tasted an aged Bordeaux, the implication was that both were superb. Because the Riojas were about $4, we purchased two whites and two reds. This contrasted with the plus $20 price tag on the cheapest of his *grand cru* Bordeauxs.

At the end of our visit downtown, we had splurged on small steaks. These we barbecued on our Hibachi situated on the balcony overlooking the parking lot. In this palatial setting, we dined in style using our borrowed lawn chairs, TV tables, finest Corelle ware, and cottage cutlery. Because of the importance of properly assessing the new wine, we took out our newly purchased official International Standards Organization (ISO) wine-tasting glasses. Thus, it was with appropriate formality that we sampled our bottle of Marqués de Cáceres red. To our joy, and a bit of surprise, the wine had an exquisite flavor and smooth taste. Our opinion of Peter's judgment took an immediate leap upward. Even after many years, when I want to give Suzanne a special joy, I bring out a bottle of Marqués de Cáceres.

During the next week–on a Wednesday to be exact–Suzanne had prepared chicken popovers for supper. Everything was ready upon my arrival from the university, so I went to select a wine from the clothes closet that doubled as our wine cellar. By accident I chose the white Marqués de Murrieta we'd picked up the weekend before. Because it was midweek, we had our regular wine glasses. I was surprised at its golden color, but took no further notice of it. After touching glasses in a toast, I lifted my glass to my nose. I was overwhelmed! What a seraphic fragrance emanated from the glass. The wine was rich, flavorful, and endowed with incredible complexity. The balance and harmony in the mouth were unbelievable. I raised my head in astonishment, only to meet Suzanne's eyes staring back at me, with what must have been the same expression of awe. I picked up the bottle again to check what I'd just opened. There it was, in gold-and-white, Marqués de Murrieta Ygay. This was no Wednesday night wine! The heavens had just opened! I had had some superb wines before, but this was *different*. It was the first time I had experienced the Holy Grail that connoisseurs seek. How could this be happening with a wine costing only $3.95, and from a region I knew next to nothing about? I'm still in rapture when I try the wine. Regrettably, the Ygay is no longer so economically priced. Even during the year the price more than doubled.

With a second great vinous experience with one of Peter's suggestions, I realized that here was a person I needed to get to know better. When I mentioned his name several days later, when talking with members of the group, everyone knew or knew of Peter. Thus, it seemed logical to ask him to discuss wine selection.

Dave's house along Triphammer Road was ideally situated as it was about equidistant from where most of us lived. The older home was also endowed with a large combination dining room/living room. It provided both a pleasant environment for discussions and had a large elongated table for tastings.

Dave's house also possessed an enviable collection of some 1,200 bottles of wine. The cellar had once been an old coal bin. However, it had undergone a wondrous transformation, with tongue-and-grove pine paneling. Dave noted that he had coated the white pine with beeswax to preserve the natural color of the wood and repel spiders. I cannot confirm whether the latter is true, but there were no cobwebs to be found! A soft wash of light bathed the bottles from fluorescent fixtures recessed behind wooden baffles. The bottles were held in a lattice of one-quarter-inch wooden doweling, inserted in 5-inch by three-quarter-inch pine paneling. The rack appeared both light and airy, and achieved considerable strength by standing the paneling upright at right angles to the wall. Positioning each pair of doweling holes every three and three-quarter-inches was ample for all standard 750 ml bottles. In some of the supports, though, Dave had placed the holes five inches apart for magnums.

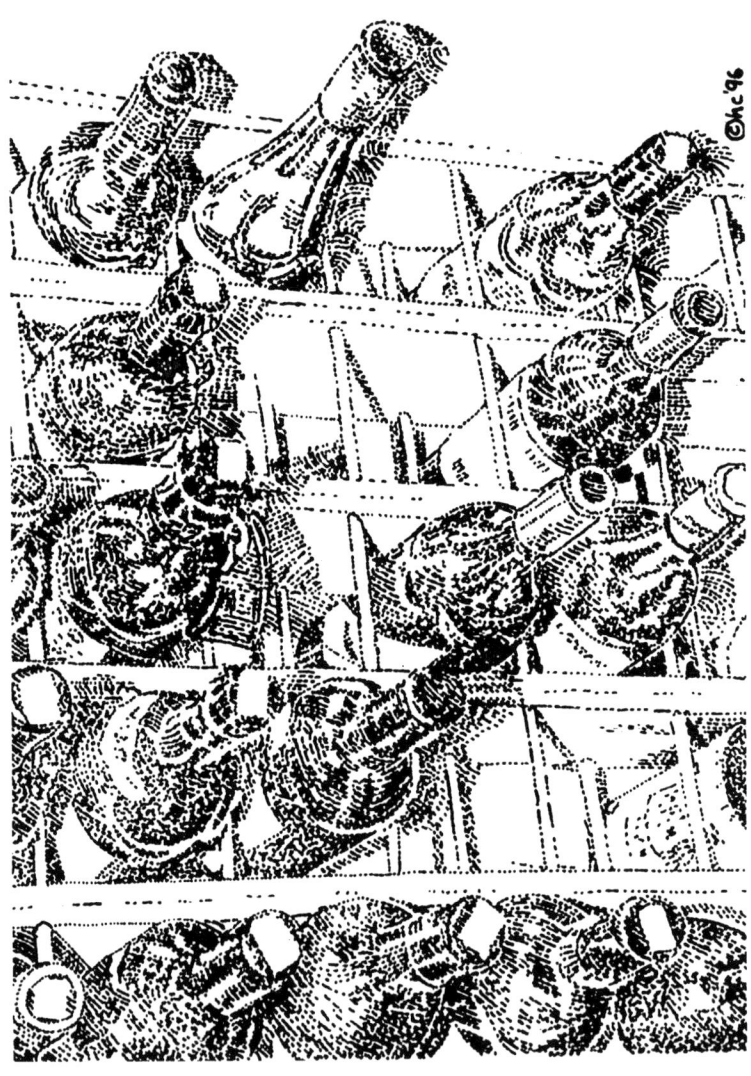

Wine Cellar Rack

78

When I arrived, Dave was in the process of introducing Peter to the members of the group. Tonight we had an additional member, Jim Minot. He was Dave's innocuous but opinionated uncle. He had unexpectedly dropped in for a visit. He liked wine but knew considerably less about the subject than he thought.

Dave explained to Peter that our meetings were very informal and we tried to prepare questions for our guest. After everyone had a glass of Niersteiner Gutes-Domtal, Dave invited Peter to begin.

TYPES OF CONSUMERS

"It's a pleasure to be able to share with you some of my ideas and experiences on wine selection. As a proprietor, I am often asked for recommendations as well as encouraged by wholesalers to promote their products. This can put me in a precarious situation, where customer service and personal profit can conflict. As far as possible, I try to forget the second and place serving the customer first. This may sound corny nowadays, but if I do a good job, the customer will probably return and my business will prosper.

"One of the first clues to a customer's wine interests is whether he or she notices the promotional material placed around the store. If the material seems invisible, I begin to suspect that the customer has a particular wine in mind . . . or has poor eyesight."

After the laughter subsided, Peter continued. "The more 'lost' people appear, the more certain I am that they are after a particular wine. In this case my help may be limited to assisting the customer in locating the wine, or to suggest an appropriate substitute.

"The connoisseur generally makes a brief survey of the store, looking first for the specialty items. They size up the quality of the store and its merchandise. For them, I may be of help with specific information about the wines or vintages available. These people, like yourselves, are a joy with whom to associate. They willingly describe the types of wine they like and usually appreciate suggestions. Assuming that the store is not particularly busy, I enjoy having long talks with these customers."

"Many stores are not blessed with owners like yourself," Dave interjected. "Normally, I try to arrange my visits when you are there. What recommendations would you give customers when knowledgeable staff is not present?"

"That's a tough question," said Peter. "It really depends on the customer. If the person knows little about wine, and simply wants something for a party, the customer is probably best led by the advertisements placed around the store. Highly advertised wines are usually of consistent quality and blended with the occasional wine drinker in mind. Blending attempts to avoid obvious flavors that might offend those unaccustomed to wines. In this regard, I must agree with the blender. Consumers new to wine are probably best to

"I don't remember the name of his favorite wine, but here's a sample."

start with off-dry neutral-flavored wines. The stronger, more asser-tive flavors that most of us appreciate are often too marked or foreign to be accepted immediately. Oddly, most periodic wine drinkers prefer very fruity sweet sparkling wines, such as Asti Spumante. In addition, they usually prefer popular brands recog-nized by their friends. The range of normal wines is far too daunting and confusing. 'Unknown' wines, no matter how good, are seldom appreciated. It is with good reason that trendy wines are prominent-ly displayed at aisle ends, or other conspicuous places in the store.

"Most people have neither the time nor the interest to study wine. In addition, many customers simply have no interest in trying different wines. Furthermore, most people don't even remember the names of wines they've tried.

"If they want an 'upper-class' wine, they're probably better off purchasing a classic European wine. Now, don't get me wrong! It's not because I think European wines are better. It's because these customers feel reassured with familiar names like Bordeaux, Bur-gundy, Chianti, Valpolicella, etc. Since most popular literature and advertisements still imply that European wines are the best (mean-ing, I assume that they are the standard by which all wines should be judged), how can one expect the average consumer to think otherwise?"

"The simpler situation comes with people like yourselves," said Peter, smiling. "Your type is seldom satisfied with mass-produced wines, nor will just any estate wine do. You generally come into the store with clear ideas of the wine you want. You also tend to be willing to experiment with different wines to expand your knowl-edge. For you, there are many ways of finding and selecting wines you'd probably like. These include many features shown on the label, such as the geographic and varietal origin, producer, vintage and, of course, price!" He laughed.

"Do you really think that these aspects are helpful?" I asked. "Take price, for example. I remember the Rioja wines you recom-mended. They were superb, *and* under four dollars each."

"I know your feeling," conceded Peter. "I added price primarily in jest. However, price occasionally can give an indication of a wine's potential quality, but only when compared with similar wines from the same region. This rarely applies when comparing wines from different countries. For example, a five-dollar wine from one country may be equivalent to a fifty-dollar bottle from another. Wines are not like cars or stereos, where price is usually a good indicator of quality. In addition, there are no industry-wide standards of quality against which wines can be measured. Each person needs to establish their own standards, and rank wines in relation to that. Too often people accept the opinions of self-acclaimed experts. Nevertheless, there are those who seem to like being told what they *should* like or buy. That reminds me of a cartoon by Mike Lynn and Bob Johnson. It shows a well-heeled individual complaining bitterly about the wine he's tasting. The store owner responds that a prominent wine critic rated the wine 96 out of 100. The customer then, without batting an eye, says, 'I'll take a case.' There is more truth in that cartoon than most of us would like to admit."

VINTAGE

"What do you think of vintage charts?" said Dave, half-laughing, as he got up to pour more wine.

"When I was younger," Peter said, "I used to follow vintage charts like others devour racing sheets or study the stock market. Now I look at them with a more jaundiced eye and only when selecting wines for my more wealthy patrons. Many of my affluent customers know little about wine, except the names of a few French estates. In addition, they accept vintage charts as gospel, and purchase only wines from great vintages. Because the sale of one case of such wine provides the profit that I make on fifteen cases of inexpensive wine, I keep my eye on the pundit's charts. They allow me to smile all the way to the bank."

"But how else are you going to avoid the terrible vintages?" retorted Jim.

"My question for you is why buy wines from countries subject to climatic vicissitudes?" Peter replied rhetorically. "Why worry about such things? If you want to avoid the problems of poor vintages, then buy wines from the regions blessed with consistently good growing conditions. When does California ever have a bad vintage, or Australia, South Africa, and Chile? I don't have to elaborate on the regions that often have poor vintages. They're the ones for which one needs a vintage chart."

"I'll not let you off that easily," returned Jim, stretching to his full five feet, four-inch height. "It's those poor years that stress the vine, so that during the good years the grapes surpass anything that can be produced from vines from soft climates like California."

"Come on, Jim! Surely you haven't fallen for that line? How could grapes going moldy a week before harvest stress the vine into producing better grapes the next year? The only stress is felt by the vintner!"

"It's a pity, Jim, that you weren't with us when we went to Venture Vineyards earlier this fall," interjected Phil, trying to maintain the peace. "We had a viticulturalist with us from Germany. He noted that stressing the vine is of value only if it directs the existing energy of the vine into fully ripening the fruit. If vines grown in consistently good climatic regions are appropriately trained, they should unfailingly produce top quality fruit."

APPELLATION CONTROL LAWS

"Well, you can't knock the Appellation Control laws in Europe," retorted Jim, sure that he was on unshakable ground at last. "They guarantee quality!"

"As a set of laws," noted Peter, "the Appellation Control regulations stipulate the grape varieties permissible and the wine production procedures permitted. In this regard, Appellation Control (AC) laws have successfully maintained the quality standards current during their introduction. However, many people, including myself, wonder if AC laws now act more as a straightjacket to progress. Technological innovation and stylistic changes are certainly more easily implemented here than in Europe.

"What AC regulations do attempt to guarantee," continued Peter, "is geographic and stylistic continuity. If this is viewed as assuring 'quality,' then I guess Jim is right. However, for me, this is not what I consider quality. In only a few countries, such as Germany, are all AC designated wines required to pass a taste test. Most or all European countries employ taste tests, but not every year, nor for all wines. Furthermore, most tests only check for faulty wines, or conformity to AC regulations. Seldom do the tests assess or grade quality. But all this aside, how does one guarantee what is not objectively definable?"

"So I suppose you think the AC system is worthless!" railed Jim.

"I did not say that, nor did I intend to imply it," responded Peter. "AC designation is not a guarantee of quality. Nevertheless, the system has its merits. As already mentioned, the AC system has maintained historically accepted norms for style and grape varietal use. For consumers, this means an element of flavor consistency one can reasonably expect from the wine. Note, though, that this does not apply to most non-European wines. The style and varietal composition of our wines here in the United States is largely up to the discretion of the winemaker. On the negative side, European AC dictates can delay the implementation of advances in grape growing and winemaking. The proof of the pudding is illustrated in Italy. There, many famous winemakers forfeit the right to use the local AC geographic name, in order to do what they consider best with their wines. Thus, many of Italy's most prestigious wines carry only the lowest AC designation."

"On a related issue, what is your opinion of the *cru classé* system?" commented Tony.

"The *cru classé* system in France is quite different in intent than the AC regulations. The *cru classé* primarily consists of a hierarchical ranking of individual vineyards. Although acknowledging that these rankings probably reflect real differences in the quality of the sites, it has definitely escalated land value, as well as increased the demand for their wines. Even consumers misuse the *cru classé* designation as a status symbol. Regrettably, this has led to the all too common, but mistaken idea, that *cru classé* wines are superior to all other wines. If the estate owners use their increased profit to improve their wines, the system can benefit at least the wealthy consumer."

"For a store owner, you've shown blemishes on sacred cows in the wine trade," I said.

"Yes, I guess I have." Peter chuckled. "It's that I see so many excellent wines that do not get recognized, and others receive far more than they deserve. That runs counter to my sense of fair play. In addition, when good producers do not receive the financial rewards they should, their ability to maintain and improve their wines is limited."

VARIETAL ORIGIN

"Thus far, you have pointed out that vintage, price, and AC designations are not necessarily trustworthy means of selecting wines. What is?" questioned Dave, as he passed around a second bottle of Niersteiner.

"You're right!" Peter said. "I have been more negative than I'd intended."

Peter pondered as he moved his hand through his locks of grey-streaked hair. He then looked up and remarked, "It must have been the dry state of my palate that affected my mind.

"Outside Europe, probably your best indicator of quality is the name of the grape variety on the label. Most non-European labels state the main grape variety used in producing the wine. Depending on the country or region, the proportion of juice coming from the indicated variety can vary. In most cases, though, the proportion is sufficient for the stated variety to dominate the wine's character. Admittedly, not all producers of a varietal wine, such as Chardonnay, make it in the same style. Nevertheless, the main variety should stamp the wine with its distinctive aroma.

"Most European wines are produced from one or a few grape varieties. Because these are seldom noted on the label, it is confusing to those just starting to enjoy wine. The major exceptions are countries with a Germanic heritage. Here, varietal designation is a long established practice. For consumers unaware of the varietal composition of different European wines, choosing non-European wines by varietal designation has much to recommend it.

"The descriptive terms found on the label are occasionally useful. An example is the term *reserve*, or its variants in different languages. Although occasionally used to excess outside Europe, most European producers are more reserved in its use, no pun intended. *Reserve* is usually employed only when the winemaker is especially proud of the wine's quality. An equivalent term, used on Portuguese wines, is *garrafeira*. They are wines worth looking for. In Germany, there are groups of producers who make especially high demands on their wines, for example the VDP [*Vereinigung Deutscher Prädikats-u. Qualitätsweinguter*] group of estates. It uses an eagle symbol on its labels or capsules. Many countries also sponsor regional, national, and international competitions. Winners of such awards are usually excellent quality wines! Regrettably, winning international prizes occasionally results in stiff price increases. The supply/demand squeeze is good for the lucky producer, but not so beneficial to the consumer's pocketbook."

GEOGRAPHIC ORIGIN

"What about the source country? Is it necessarily a good indicator of quality?" asked Claude.

"Yes," said Peter hesitantly. "But you have to take most of what you've read or heard with a large grain of salt. Advertisement is advertisement, and can be biased, as you might expect. Wine writers try to be honest, but may have to sacrifice complete objectivity to satisfy their employers. Remember that publishers earn most of their revenue from ads. Repeated bad press could lead to advertisement being pulled. In addition, if you had your visit to a country, region, or producer paid by the same, would you file a negative report about their wines? Maybe, but if you did, your free trips would disappear like water in the Sahara. The only ones you may be able to trust are those who have nothing to gain by saying the truth. On second thought, there may already be too many people like that. Just reflect on some of the unsolicited opinions you may have heard at wine tastings. It tends to be proprietors, like myself, who try to steer people toward good wine at reasonable prices. That is, if we sense that this is what the customer wants."

"May we ask you for your own preferred wine regions?" requested Tony.

"Since my biases are no secret, sure. I hope no one will be offended if I miss their favorite region. My tastes are broad, but not all-inclusive.

"For soft, balanced, perfumed white wines, my preference still leans to Germany, especially from the Mosel and Rheingau. They have the character and aroma I love. If I knew nothing about the producer or site, I would select a bottle without English on the label–in other words, a label designed for the domestic German market. Exported wines of this type are generally of better quality than blended wines, which are appropriately labeled for export. The major German exporters are not unaware that anglophones have difficulty with Teutonic names.

"For dry white wines, my preference list is more extensive. For complex whites, I have been especially pleased of late with Australian wines. Californian Chardonnays also rate highly in my book. Our own local wines can be excellent, but tend, in my humble opinion, to be better without exposure to oak. I am quite pleased with our Gewürztraminers, but for the variety's steely tension I like, versions from Alsace still have the edge in my view. Pinot Grigio wines from northern Italy are often pleasantly but subtly distinctive. Sauvignon blanc wines are not among my favorites. They too often exhibit a herbaceous, bell pepper aspect that is not to my liking. For something special, I often select a white wine from one of the traditional producers in Rioja.

"If looking for an excellent example of an aged red wine, Rioja is again one of my favorites. Their Reserva and Gran Reserva wines are usually excellent buys. Alternately, an old garrafeira from Portugal is almost assured to please. For more fruity to jammy red wines, one has a wide range of excellent Californian and Australian products at competitive prices. Both areas also produce an interesting variety of up-scale reds, largely from Cabernet Sauvignon and Syrah (Shiraz in Australia). For Pinot noir (Burgundy-like) wines, I would prefer to put my money on some of the recent wines produced in Oregon. Claude, here, often has extolled the excellence of their wines. Burgundy produces some excellent Pinot noir wines, but they tend to be expensive. I still stock a few Burgundy's from producers I trust, for those who insist on having them. Italy is producing excellent wines from French varieties, but my preference is for wines from their own indigenous cultivars. Here, though, you can't go by varietal name. This is because few people can distinguish the names of Italian grape varieties from producer or estate names. Although some of these varieties produce uninspiring wine, others are a real revelation. They can possess great flavors that most of us have never imagined could occur in wine. The super wines made from the 'Negro amaro' and 'Malvasia nero' grapes are one of my recent discoveries in this regard. Tony helped me latch on to these varieties. He also pointed out an excellent Italian varietal wine (Barbera) produced by Santino in California. I'm always looking for great new wines for my customers.

"Finally, when it comes to sparkling wines, I'm afraid that I am frightfully conservative. I do prefer the wines from that celebrated region in France. Champagnes have a toasty fragrance that, combined with their subtle balanced flavors, is almost sinfully ravishing. Concerning sherries and ports, I'm also dreadfully traditional. I prefer those produced in the countries where the styles developed–Spain and Portugal."

AGING POTENTIAL INDICATORS

"One aspect that interests us, but you haven't mentioned as yet, is how long we should age wines," noted Phil.

"Ah! . . . The question of aging potential," said Peter, as he rested his head on his upraised left hand. He seemed to be searching for inspiration. After a few seconds, he cocked his head and began. "If the wine does not mention a vintage date, then the wine probably will not improve much with aging. Regrettably, the reverse is not necessarily true. Most vintage-dated wines do not age well either. Just think of Beaujolais nouveaus. Most white wines do not improve significantly after bottling, although there are several important exceptions. Mature, oaked, white wines often do improve, or at least the changes that occur are appreciated. However, it is usually red wines that one gets questions about aging potential. With European wines, this is usually related to weather conditions during the vintage."

There was a pause, as if a new idea had suddenly surfaced from the recesses of his subconscious. "This may sound strange, but I would look at the length of the cork in the bottle."

"Length of the cork?" exclaimed Pat, looking perplexed. "What's that got to do with aging potential?"

"First of all, it often is an indicator of the winemaker's opinion concerning the aging potential of the wine. The best premium wines are stoppered with two-inch corks. Medium quality wines, with several years of aging potential, generally have one and three-quarter-inch long corks. For wines with little aging potential, one and a half-inch corks usually suffice. Although cork length does not translate directly into the number of years the wine will improve, the best quality long corks are usually reserved for wines that need the oxidation protection long corks provide.

"The next time you remove a cork from the bottle, note its length and quality. By quality, I mean freedom from obvious cracks, cavities, and other physical imperfections. I am sure you'll soon recognize the connection between cork length and wine quality. While it's not possible to detect the quality of cork in an unopened bottle, it's often easy to determine the cork's length."

"What should I do if I open a bottle of red wine too early, and find that it is too bitter and astringent?" I asked.

"It depends on the conditions under which you open the bottle. If it's at a party, there's not much you can do. However, you ideally should open the bottle to sample the contents before serving it to your guests. That way, you can avoid the embarrassment of presenting a faulty wine, or you can recork the bottle if it's still too young. Leaving the latter for several days may result in a miraculous transformation. The small amounts of oxygen absorbed during opening may induce tannins to interact. The resulting complexes are less bitter and astringent than the original tannins. *Voilà*, a smoother, more pleasant wine."

"What happens if you've already sampled too much wine to refill the bottle?" I continued. "How much empty space can you safely leave in a recorked bottle?"

"You don't want to leave the bottle any less than 90 percent full," responded Peter. "More than that and the wine may oxidize excessively and lose its desirable flavor. Also, the wine may turn vinegary in several months.

"I also use half or quarter screw-cap bottles to temporarily store wine. They are ideal for partially consumed bottles. In fact, since my wife and I seldom drink a whole bottle with supper, I pour half of its contents into a half bottle before sitting down to eat. Filled to the top and tightly sealed, the wine remains in perfect condition for days."

BAG-IN-BOX

"An alternative solution to the problem of oxidation is the use of bag-in-box wine. Although not classy, nor containing the best wine, the bag-in-box technology has distinct advantages. This is especially so for single people. You simply take the quantity you want, and the rest remains protected from air."

"I'm not familiar with bag-in-box wine." said Jim. "How do they avoid the oxidation problem?"

"The pressed carton box contains a specially designed, double-lined, metallized plastic bag that is filled under vacuum. When the tap is pressed, wine flows out as atmospheric pressure collapses the flexible bag. In contrast, air must enter a wine bottle for its contents to flow out. Thus, wine can be periodically withdrawn over several weeks from bag-in-box containers without the wine going off."

"If the technique is so good, why don't producers only use such containers?" questioned Jim again.

"Tradition," responded Peter half jokingly. "Well, not exactly. There are problems with the technique, notably with the junction between the tap and the bag. There is a tendency for the wine to slowly absorb oxygen and oxidize. Thus, the technology is appropriate only for wines that do not need aging and are likely to be consumed shortly after reaching the market.

"But now, back to indicators of aging potential! One of the best indicators is the grape variety. Red wines from varieties like 'Nebbiolo,' 'Tempranillo,' 'Cabernet Sauvignon,' 'Syrah,' and 'Pinot noir' generally age well. White varieties, producing wines with substantial aging potential, are 'Riesling,' 'Chardonnay,' 'Sauvignon Blanc' and 'Viura.' This list is, of course, incomplete, but may be of help. The actual aging potential of a wine depends considerably on the quality of the harvested grapes, the skill of the winemaker, and of course the conditions of storage."

"This may seem simplistic, but I'm still not perfectly clear on what people mean by aging potential," said Pat.

"You're not alone! Many people have the impression that all wines improve during aging, eventually reaching a climax, and then fall apart, becoming vinegary. In fact, most wines are near their best when released from the winery. They maintain their qualities for six months up to several years, and then progressively lose their initial flavor. In contrast, premium quality wines generally give greater pleasure–meaning less bitterness and astringency–if stored for several years. During aging, the initially fruity character may increase, taking on a jammy note. With further aging, the wine's varietal aroma fades and is replaced by an aged bouquet. In superb wines, the latter often provides more delight than the original varietal aroma. The longer these changes give pleasure, the longer is the wine's aging potential. Thus, it's better to speak of a broad plateau, rather than an apex, at which the wine is best. When the changes result in the wine losing its pleasurable character, it's considered 'over-the-hill.' In the end, the wine will just be a flimsy shadow of its former self.

"One further point–different critics have markedly different views on aging potential. It's fascinating to compare the aging predictions of French, British, and American pundits, for example, on Bordeaux wines. French commentators often give values ten or more years earlier than their British counterparts. American critics tend to fall somewhere in the middle, between French impatience and British reserve. That aging potential is not precise is part of the intrigue of opening older wines. One is both eager and apprehensive. Is one likely to commit infanticide, open a bottle of earthly paradise, or release the ghost of what once was?"

"Do old wines really turn into vinegar?" I inquired.

"No, at least to my knowledge. I haven't tried many wines over fifty years old. Nonetheless, those I've tried were flabby undistinguished liquids with hardly any bouquet left. If the cork fails, air and bacteria may enter the bottle, turning the wine to vinegar."

"If you don't mind my interrupting," Dave remarked, "I think it's time to let Peter have a break. Although Peter hasn't mentioned it tonight, I know he has a passion for dessert wines from South Africa. I just happen to have a bottle of Nederburg Edelkeur. I thought that Peter, and all of you, might enjoy trying it. I also have a small passionate treat that might go well with the wine: rich, creamy, vanilla ice cream smothered with passion fruit! A little decadence is good for one, occasionally."

Peter's eyes opened wide. "Wow! I haven't tried a bottle of that elysium since my trip to Cape Province. What a treasure!–and at such a fitting moment. I had said all–no, more–than I had intended."

As Dave poured the liquid gold into small crystal glasses, we got up to individually thank Peter for his fascinating presentation. None of us, except Dave, had thought of bringing a token of our appreciation. Thankfully, Dave's presentation of the Edelkeur seemed adequate.

LABEL REMOVAL

As we mingled around, I asked Peter if he knew of a good way to remove wine labels. Suzanne and I wanted to place the label next to our notes in a wine log we'd recently purchased. Soaking in hot water often worked, but not always. Peter suggested trying some household ammonia or strong detergent. For labels that were still recalcitrant, he suggested using a single-edged razor blade to slowly slice the label off the bottle, after prolonged soaking. Because most of the glue remained on the label, it was necessary to position the label on wax paper or plastic vegetable wrap before pressing.

Peter asked me how I intended to press the labels. At this, I related my use of a herbarium plant press. After patting the labels semidry on a towel, I'd place them between sheets of felt paper, separated by sections of corrugated cardboard. The collection was placed between slabs of plywood, with the bucket of water–involved in soaking off the labels–used to supply weight. Within a day, the labels were dry and looked almost brand-new. If not, pressure from an iron removed most creases. Peter liked the technique, as it seemed more effective than his use of newspapers and heavy books.

After enjoying the presentation and wine, it was hard to pull myself away. However, I wanted to review my talk on sclerotial physiology I was to give the next morning. So with considerable regret, I had to bid my farewell and go to work.

Chapter 6

Bacchanalian Pleasures and Faults

None of us, except Phil, had training in sensory evaluation. To improve our skill, we had asked Craig Goodwin to instruct us in the finer points of wine degustation.

Craig was a wine journalist, titular head of the Bacchus Society, and taught wine appreciation courses through Cornell's Extension Branch. I had met him briefly at the Bacchus Society, but hadn't had a chance to get to know him. He seemed quite a character–bursting with energy and awash with ideas. He seemed ideally suited to guide us through the intricacies of wine tasting. All our other members knew Craig, either from the Bacchus Society, or from his newspaper column.

When Phil phoned Craig, he asked us to bring any faulty wine we might have. It sounded strange, but we did our best. Phil had the largest selection. He had asked Peter Maples for any bottles unsatisfied clients had returned. I, myself, had nothing to contribute. Together, the group had seven bottles of faulty wine. We also brought good wine, for what we assumed would be the main component of the tasting.

WINE GLASSES

For a change, all of us arrived early. Thus, when we heard a car door slam and footsteps on the veranda we were sure it must be Saint Nick–no!–Craig was here. Dave went to the door, and there, faintly silhouetted against the fading sunset, was our curly-haired Einstein of bacchanalian pleasures. His enthusiasm was irrepressible. People loved his energetic and direct delivery.

After casting his coat on the corner rack near the door, he surged in with his six-pack of clinking glasses. Toting wine glasses in a

Coke carton was a telltale sign that one had taken Craig's course. The small-sized carton was perfectly suited to housing the Libby® #8470 glasses. These were standard equipment in his course and for the Bacchus Society. Dave directed Craig to the head of the table, while the rest of us took what were becoming our regular places.

Planting his wine glasses in a semicircle in front of him, he commented, "It's not without reason that my motto is '*Have glasses, will travel.*' Although I use various glasses for different wines at home, for the serious matter of tasting I use the ISO wine-tasting glass. It was designed by the people who brought us temperature in Celsius, pressure in Pascals, and wine volume in Hectolitres.

"Note the distinct tulip shape of the glass. The tall, sloping sides help you avoid anointing your neighbors while swirling the wine. This is also a good reason why you should never fill your glass more than one-third full. It may be a fetish, but I become irate when waiters fill my glass to the rim. I could drown when plunging my nose in to savor the wine's delights. Also, I hate coming up dripping."

Dave interjected, his eyes twinkling. "I assume that my wife's favorite lead-crystal water goblets are not suitable for our deliberations?"

Craig cracked back, "You're darn tootin'! Those containers may be fine for insipid fluids like water, but *not* wine. I do approve of regular crystal, though."

At that same moment he tapped two of his glasses together, producing a beautifully resonating *ping*, denoting their crystal nature.

"Even more ridiculous are the short-stemmed fruit cups misused as vessels for sparkling wine. How can one appreciate the joyous effervescence of a champagne with anything other than a slim crystal flute?!" asserted Craig. "You pay a hefty luxury tax for those bubbles; you might as well see them!"

"Other than the pretty sound, does it really make any difference if the glass is crystal?" I questioned.

"Ask a person with a BMW whether it makes any difference in getting from point A to point B. Regular glass can do the job as well, but the light delicate feel of thin crystal, combined with the joyous resonance, infuses refinement into every vinous event.

ISO Wine-Tasting Glass

"Glasses should also be colorless and plain. Colored glass distorts the wine's hue and its interpretation. Attractive as decorative etching may be, this detracts from the appreciation of the wine's appearance.

"There's another reason for bringing my own glasses, other than demonstrating the ISO glass." Craig laughed. "I know they've been properly rinsed and stored."

At the same time, Craig ducked, as if Dave were about to throw something at him. "No, Dave, I know your glasses are cleaned properly. However, I've been in restaurants where improperly washed glasses have ruined the wine's fragrance. In addition, storage in a painted cupboard can taint the glasses. As a matter of habit, I now smell every glass before using them. If unacceptable, I order replacements. It is crucial that wine glasses be as odorless as they are spotless."

"May I interject for a moment?" questioned Pat. "I've noticed that wine experts often hold their glasses in an odd manner–by the base. Is there any rationale for this procedure?"

"Yes! . . . pure unadulterated exhibitionism! The posture is as contrived as pouring a bottle of champagne with your thumb in its base. Nevertheless, holding your glass by its stem makes sense. Often people think this is important to avoid heating the wine. Although true in an absolute sense, holding the glass by the bowl warms the wine only insignificantly. Using the stem does, though, avoid soiling the bowl with fingerprints."

"In your example about holding champagne bottles," replied Pat, "you noted an interesting feature. Does the punt in wine bottles really help in collecting sediment, as I've read?"

"I don't know whether the punt helps collect sediment, but it does reflect the evolution of bottle manufacturing. As you probably know, bottles were originally handblown. Large scissorlike utensils were used to cut the bottle free from the metal blowpipe. To hold the bottle during this process, a metal pontil rod was attached to the bottom using a dollop of molten glass. Pushing the pontil rod into the still soft base of the bottle produced a flat rim, helping the bottle to stand upright. The procedure also buried the sharp edges of glass produced when the pontil rod was snapped off. Subsequently, enterprising scoundrels used punts to surreptitiously reduce bottle volume."

Since there were no further questions, Dave rose to get the first of our white wines. He had placed them on the patio when we had arrived. With temperatures falling below freezing every night, the wines were sure to be cool. Dave's departure was our clue to bring out our glasses. Luckily we had all brought our Bacchus Society glasses. Although not crystal, they were less fragile and had an appropriate tulip shape.

CORKSCREWS

"The next requirement for a serious taster is to have an appropriate selection of cork-extracting devices." At this, Craig opened a box containing a fascinating assortment of corkscrews and related paraphernalia. It really was an example of human ingenuity in action. The first out of his treasure trove was the chrome-plated, wing-handled corkscrew that must abide in almost every household across the nation. He held the device up with two fingers, as if it were a smelly fish. "This is an abomination!" he said. "Its solid screw can do mortal injury to all but the youngest corks." Then he lifted out of his coffer his favorite corkscrew, the Screwpull®, as if it were a jewelled necklace. "Its hollow-centered, Teflon-coated screw glides through cork as if it were butter."

I was amazed to watch Craig extract the cork from the bottle he had in his hand. In the same continuous twisting action, the corkscrew effortlessly traversed and extracted the cork. No Herculean effort was needed to yank the cork out of a bottle clenched between knock-kneed legs. What a godsend for servers obliged to wrestle reluctant corks out of protesting bottles.

I wanted to ask where one could purchase a Screwpull, but Craig was already demonstrating what he called the Ah-so. It was a nifty combination of two flexible steel flanges affixed to a handle. The flanges of different length were independently inserted just between the cork and glass, and then seesawed down the neck. A combination of twisting and pulling cleanly extracted the cork from its housing. Craig remarked that the Ah-so was particularly useful with corks that had been in the bottle for twenty years or more. "There is a tendency for old corks to split, even with careful use of my trusty Screwpull," he said.

Craig also demonstrated the knot-on-a-cord technique of extracting corks which were inadvertently pushed into the bottle. Lowering the knotted end of the cord into an empty bottle, he acted out how gentle pulling might result in the knot catching under the cork. With considerable effort, such a perverse cork could be removed. Obviously easier on the hands was a device with six flexible plastic flanges, possessing backward directed barbs on their outer surfaces. When inverted and inserted into the bottle, the flanges flare out. On pulling, the flanges close on the cork, and the teeth bite into the stopper. Slow pulling can easily extract a deviant cork.

Corkscrews

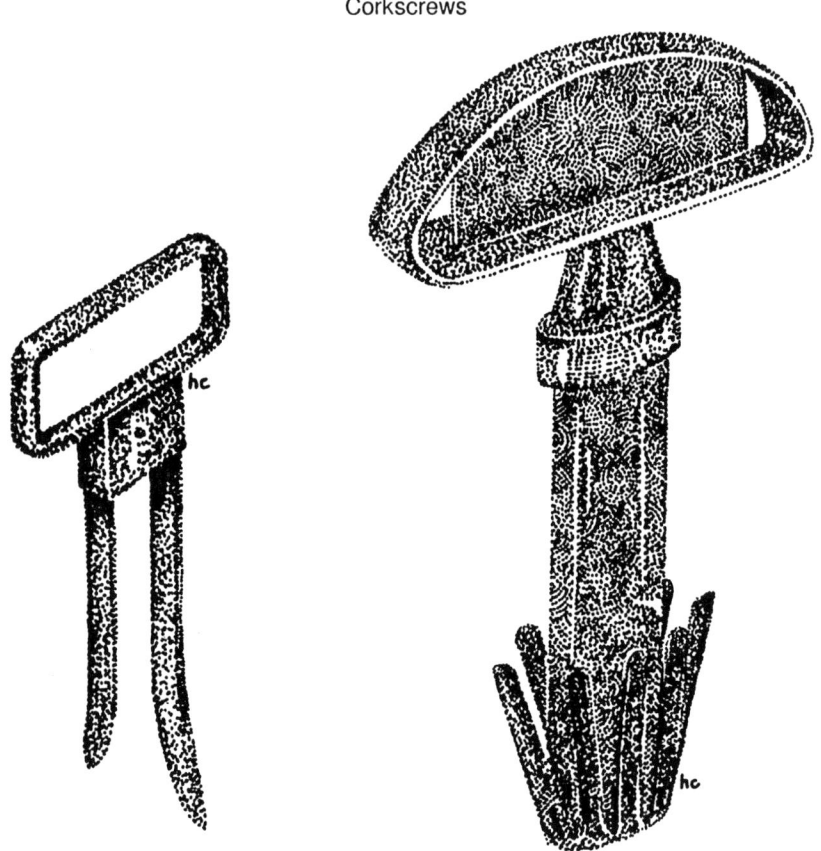

WINE APPEARANCE

Our first wine of the evening was a Cayuga blanc from Glenora Wine Cellars on Seneca Lake. 'Cayuga' was a new cultivar to me. After each of us had poured a third of a glass, Craig started to discuss the wine's appearance.

"Notice the pale straw color of the wine. Its color reflects its young age and absence of prolonged skin contact."

At this, Claude interjected. "The winemaker at Heron Hill mentioned skin contact when we were at the winery several weeks ago. However, would you explain the meaning of the term again?"

"Skin contact refers to the time the juice is left in contact with the seeds and skins, between grape crushing and the start of fermentation. In our Cayuga blanc, essentially all the color comes from the juice, without significant pigment pickup.

"Knowing the age and style of wine usually allows one to predict its color. If it differs significantly from expected, it suggests the possibility of a fault. Of these I shall talk later."

"Of what significance are the bubbles that have formed?" asked Pat, pointing to the tiny spheres that had developed on the bottom of his glass.

"Very little," was Craig's response. "In most cases it signifies only that the wine was supersaturated with carbon dioxide when bottled. Wines bottled shortly after fermentation often contain excess carbon dioxide. This subsequently escapes in the bottle. Rarely, a slight fizz may also occur due to a second in-bottle fermentation. Depending on the causal organism, the wine may or may not be cloudy, and possess taste and odor faults.

"This conveniently leads us into visual faults," continued Craig. "Probably wines are rejected more because of the presence of tartrate crystals than any other reason. Consumers too often misinterpret these harmless, tasteless crystals of cream of tartar as splinters of glass. They typically develop in wine that is cooled excessively. The Germans have a congenial term for these crystals—'wine diamonds.' The shimmer of light reflecting from these transparent particles can be enchanting.

"Suspension of sediment in the bottle is another potential cause of haziness. However, in contrast to tartrate crystals, suspended sediment seldom induces wine rejection. Sediment typically occurs only in older wine, seldom available in restaurants where wine rejection is the most common. Also, people owning or ordering such bottles probably understand the origin of sediment. In addition, bottles of aged wine are typically handled gingerly, and the liquid decanted slowly to separate the wine from any sediment. Some connoisseurs even expect to find sediment in their bottles. It's taken as a sign of quality—an indicator that the wine was not overly fined."

"When we were at Heron Hill, we heard of racking to remove young wine from its sediment. How come sediment still forms after racking?" inquired Claude.

"Wineries may use several procedures to remove sediment, notably racking, as you mentioned, as well as fining, filtering, centrifuging, etc. However, as wine ages–especially red wines–pigments, tannins, and dissolved proteins form complexes. These may combine with tartrate crystals and precipitate, forming what is called sediment. The more a wine is fined, the less likely it will produce sediment in the bottle."

"You noted that some people consider the presence of sediment a quality feature. Of what benefit is it?" I asked.

"It's not that the sediment is of value," responded Craig. "However, too much clarification may unintentionally remove vital flavorants from the wine. There is evidence that this belief is based on fact, and not simply marketing caprice by premium wine producers."

"I recently had an argument with a friend of mine," commented Claude. "He said that he had had a wine in which sediment was suspended, but that it had no bitter taste. I didn't think that was possible. What's your opinion?"

"Sediment often tastes bitter, as most of you know, but not always. Occasionally, it's more chalky and gritty than bitter," responded Craig.

WINE FRAGRANCE

Having briefly commented on visual clues, Craig began to discuss wine fragrance. "Taste may be the heart of a wine," Craig noted, "but the fragrance is its soul. The sensation of a superb fragrance is transcendental. However, experts differ on how best to assess this property.

"One of my *confrères* in Chicago prefers to begin by wafting his hand over the glass toward his nose. Although somewhat histrionic, there is logic in the procedure. It permits you to detect separately the wine's most volatile aromatic compounds. Often I tend to be too impatient and want to plunge directly into the glass and inhale the full richness of the fragrance. Swirling the wine in the glass helps to release aromatic compounds. Its sloping sides further concentrate the aromatics as they rise to the mouth of the glass."

At this point, Craig handed out two Wine Aroma Tables, developed by one of his friends. "Although informative," he com-

White Wine Aroma Table

Wine Aroma Table: *White Wine*

VARIETY (Region)	Flavor	Sweetness	Oakiness	Aging Potential
o CHARDONNAY (Burgundy, F) Apple, melon, peach, buttery, almond	● ●● ●●●	dry	+/-	○ ○○ ●●●
o CHENIN BLANC (Sancerre, Loire, F) Camellia, guava, waxy	○○ ○○○	dry/sweet	-	○ ●● ○○○
o GARGANEGA (Soave, I) Fruity, almond	● ○○ ○○○	dry	-	● ○○ ○○○
o GEWURZTRAMINER (Alsace, F) Litchi, Citronella, spicy	○ ●● ●●●	dry/semidry	-	● ○○ ○○○
o MUSCAT (Greece) Grapy	○ ○○ ●●●	dry/semidry	-	● ○○ ○○○
o PARELLADA (Catalonia, S) Apple, citrus, cinnamon, licorice	● ●● ○○○	dry	-	○ ●● ○○○
o PINOT BLANC/BIANCO (Alsace, F - Veneto, I) Fruity, romano cheese	○ ●● ○○○	dry	-	● ○○ ○○○
o PINOT GRIS/GRIGIO (Alsace, F - Veneto, I) Passion fruit	● ○○ ○○○	dry	-	○ ●● ○○○
o RIESLING (Mosel, Rheingau, G) Rose, fruity, pine	○ ●● ○○○	dry/sweet	-	○ ○○ ●●●
o ROUSANNE (St. Joseph, Rhône, F) Peach	● ○○ ○○○	dry	-	○ ●● ○○○
o SAUVIGNON BLANC (Bordeaux, F) Bell pepper, olive, herbaceous, floral	● ●● ○○○	dry	+/-	○ ●● ○○○
o SEMILLON (Bordeaux, F) Figs, melon	● ●● ○○○	dry/sweet	+/-	○ ●● ○○○
o TORBATO (Sardinia, I) Green apple	● ○○ ○○○	dry	-	● ○○ ○○○
o VIURA (Rioja, S) Vanilla, butterscotch, banana	● ●● ○○○	dry	+	○ ●● ○○○
o BOTRYTIZED Wine (Sauternes, F - Rheingau, G) Apricot	○ ○○ ●●●	sweet	-	○ ○○ ●●●

Red Wine Aroma Table

Wine Aroma Table: *Red Wine*

VARIETY (Region)	Flavor	Oakiness	Aging Potential
o ALEATICO (Apulia, I) Cherries, violets, spice	o ●● ooo	+	● oo ooo
o BARBERA (Piedmont, I) Berry jam	● ●● ooo	-	● ●● ooo
o CABERNET FRANC (Bordeaux, F) Bell pepper	● oo ooo	+/-	● ●● ooo
o CABERNET SAUVIGNON (Bordeaux, F) Black currant, bell pepper, leather, cigar box	o oo ●●●	+	o oo ●●●
o CORVINA (Amarone, I) Tulip, daffodil, oxidized, spicy	o ●● ooo	-	o ●● ooo
o DOLCETTO (Piedmont, I) Quince, almond	● oo ooo	-	● oo ooo
o GAMAY NOIR (Beaujolais, F) Kirsch, raspberry, fruity	o ●● ooo	-	● oo ooo
o GRIGNOLINO (Piedmont, I) Clove	● oo ooo	-	● oo ooo
o MERLOT (Bordeaux, F) Black currant	o ●● ooo	+	o ●● ooo
o NEBBIOLO (Barolo, Barbaresco, I) Violet, rose, truffle, tar	o oo ●●●	+/-	o oo ●●●
o NERELLO MASCALEA (Apulia, I) Violet	o ●● ooo	-	o ●● ooo
o PINOT NOIR (Burgundy, F) Cherry, beet, mint, ham	o ●● ooo	-	o ●● ooo
o SANGIOVESE (Chianti, I) Cherry, violet, licorice	o ●● ooo	+/-	o ●● ooo
o SYRAH/SHIRAZ (Chateauneuf-du-Pape, F) Currant, violet, berry jam, pepper	o oo ●●●	+/-	o ●● ●●●
o TEMPRANILLO (Rioja, S) Citrus, incense, berry jam, truffle	o ●● ooo	+	o ●● ooo
o TOURIGA NACIONAL (Dão, P) Cherry, mint	o ●● ooo	-	o ●● ●●●
o ZINFANDEL (California, U.S.A.) Raspberry, berry jam, pepper	o ●● ooo	+/-	o ●● ooo

mented, "don't let the tables make you think that identifying wine fragrances is necessary for appreciation. Once, I thought so. However, in what other food do we describe its flavor in terms of other products? Do we describe the aroma of peaches and apples relative to lemons, beets, mushrooms, tar, fusel oils, or cigar boxes? I now prefer to describe wine fragrance in terms of the richness and complexity of its varietal or stylistic nature. Aroma descriptors are justified only when the similarity is striking, as for example with the litchi nut aroma of Gewürztraminer wines, or the walnut aspect of some sherries. The primary value of the Aroma Tables is in directing your attention to the wine's fragrance, something too few people do. By focusing on the sense of smell, you fix your thoughts on the wine's most complex and pleasurable aspects.

"Although useful, don't permit the exercise to degenerate into a parlor game of identifying grape varieties, geographic origin, or vintage date. I'm sure you've been annoyed by individuals so engrossed in wine's statistics that they don't even seem to enjoy the wine. It's like describing a meal in terms of its components, spices, and preparation procedures. Wines, like meals, are to be savored, not dissected. Sensory analysis is crucial only for professionals attempting to improve the quality of the wine, or discern the origin of faults. Wines occasionally may rise to the level of an art form, but they are to be consumed, not placed on pedestals or hung on the wall."

"If we are not to be overly concerned about naming varietal fragrance or geographic origin, what *are* we to look for?" questioned Pat.

"I wish it were easy to explain," Craig said with a sigh. "It's like trying to explain the difference between good and sublime poetry. Most poetry is eminently forgettable, but inspired poetry has elements that can make it divine. Like poetry, most wines don't benefit from close scrutiny. Focusing too heavily on its fragrance may only highlight its limited aromatic endowment. For great wines, serious study directs your attention to aspects such as complexity, harmony, balance, development, duration, and sustained interest.

"Complexity denotes the presence of many separate aromatic elements, rather than one or a few easily recognizable odors. Harmony refers to the balance of fragrances and tastes. They produce an overall favorable response in which no single odor dominates. Development designates changes in the aromatic character that occur during tasting. Ideally, these changes maintain your interest and keep drawing you back to discover the wine's latest transformation. Duration indicates how long the fragrance retains a unique character, before becoming simply winy. Interest is the combined influence of the previous factors on the taster's attention. If the overall experience is sufficiently marked, your experience is raised to the plane of memorableness. This is the apotheosis for which we all search!"

"Although it's seldom happened to me, I can affirm its intensity when present," said Dave. "It's like a chalice of the purest joy, an out-of-body sensation."

"Until a person has had the experience, one can't understand the feeling," reflected Craig. "Although the perception may be short-lived, its memory trace lasts forever! However, let's get back to earth.

"When smelling wine, you should take a normal whiff. There seems little value in taking artificially deep breaths. Furthermore, prolonged inhalation should be avoided. Our olfactory receptors become rapidly adapted to most odors. This is why the fragrance seems to fade when wine is smelled too long, or at too frequent intervals. It may take about a minute or so for the receptors in the nose to reestablish their normal sensitivity. The only time I've found prolonged inhalation to be of value is with Vintage port. Its fragrance is so complex that, as the nose becomes adapted to some aromatic compounds, others appear as if being unmasked. I tell my students it's like viewing a panorama and progressively focusing into the distance. Well, I'll leave you free for a few minutes to concentrate on the fragrance of the wine. Remember now . . . just the fragrance!"

After we had sampled the Cayuga blanc, Craig said, "What do you think?"

For a few moments there was no response. For once, I decided to speak first. "Although a bit fruity, there doesn't seem to be much. Am I missing something obvious?"

"Any other comments?" asked Craig, looking around the table.

"There may be a hint of film developer," said Claude. "But then, that may be because of my experience in radiology."

"Does anyone other than myself detect elements of *porcini* mushrooms in the wine?" offered Tony.

"What on earth are porcini mushrooms?" blurted Pat.

Craig laughed. "Porcinis are the celebrated Italian mushrooms that cost a fortune. But your responses are interesting in their individuality. It's not uncommon for wines to provoke such a divergence of opinion. Cayuga blanc does not possess a particularly marked varietal aroma. Thus, when asked to describe the fragrance, people search for terms familiar to what they detect. This phenomenon is not restricted to those unfamiliar with wine. Wine critics commonly express this tendency. The terms used often reflect the cultural and geographic background of the critic. Note the frequent reference to truffles, cherries, and irises in northern Italian wines; black currants, peaches, apples, and melons in some French wines; and the perfume of roses and other blossoms in German wines. It's fun to search for identifiable aromas in wine, but remember that it may tell more about the individual's experiences than it reflects the fragrance of the wine.

"There may also be another source for the elaborate wine descriptions occasionally seen—the need to supply copy to editors. Don't quote me on this though. When pressed for time and short of ideas, verbose sensory descriptions help fill the creative void."

WINE TASTE

"Now, let's progress to actually tasting the wine. There's considerable diversity in the tasting procedure recommended by different authorities. The appropriate procedure also depends on the purpose of the tasting. On occasions like this, I prefer to do it in a precise manner, where I look for each taste and touch sensation, and their mutual interaction, in sequence. However, I do not recommend this approach at the dinner table!

"As soon as you take a sample, rotate the wine in the mouth so that it contacts all your taste receptors—what people call 'chewing' the wine. Concentrate on the sequence and intensity of the taste sensations found in wine: sweetness, sourness, and bitterness. Concentrate also on the so-called mouth-feel sensations: astringency (dry, dusty, puckery), heat, weight, smoothness (absence of astringency plus viscosity), etc.

"When present, sweetness is the first taste sensation detected. Sweetness is also the first sensation to fade. If weak, its presence is soon masked by other more intense and persistent tastes. Sweetness is most quickly and readily detected on the tip of the tongue.

"Sourness is also rapidly detected. Depending on the individual, wine acidity may be sensed most markedly on the sides or under surface of the tongue, or on the cheeks. If the wine is sufficiently acidic, it also stimulates a dry, puckery, astringent sensation. Occasionally, people detect saltiness in acid solutions. This probably results from the stimulation of salt receptors by acids. Although tartrate salts occur in wine, they do not produce a salty taste.

"Bitterness, if detected, usually becomes evident within five to ten seconds. Its perception can last for several minutes after swallowing the sample. That is, assuming you down the wine, rather than spitting it out," said Craig, grinning.

This comment certainly got everyone's attention. There was no intention of having the wine go in any direction other than that for which it was intended.

"Most bitter compounds are detected at the back, central portion of the tongue. However, some bitter compounds are best detected on the tip of the tongue. The tannins found in red wines produce a generalized, dry, dusty, puckery, astringent sensation throughout the mouth. Its detection is usually slow, often taking seven to fifteen seconds to reach peak intensity. The sensation is also slow to fade. It is one of the main reasons why tasters may take bread between samples to remove the sensation of the previous wine."

"I have heard the expression 'Sell wine over cheese, but buy it over water.' Can you explain that saying?" asked Tony.

"The person who coined the expression knew well how food influences wine perception! The reason cheese is served at wine tastings is to diminish the bitter and astringent sensations[1] often found in many young wines. However, if you want an unbiased perception, it's better to use water as a neutral palate cleanser. An even more marked example of taste modification follows tooth brushing. The sodium lauryl sulfate found in most toothpastes temporarily disrupts the cell membrane of the taste receptors."

"What causes the dry, dust-in-the-mouth sensation of tannins?" I inquired.

---

[1] caused by the presence of salt in the cheese

"Apparently tannins react with saliva proteins, causing them to settle on surfaces of the teeth and mouth. On the teeth, the precipitated proteins generate the filmy coating detected by the tongue. On the lining of the mouth, the precipitated proteins force water away from the cells. It is the latter that gives the dry sensation."

"Does any of this relate to why different wines are served at specific temperatures?" I continued.

"Partially," responded Craig. "Cool temperatures augment the bitter and astringent sensations of tannins; therefore, there is a preference for red wines at room temperature. Conversely, cool temperatures suppress the sense of sweetness. This minimizes the cloying sensation of very sweet wines. However, for dry white wines, cool temperatures

"They gotta be kidding: Serve at room temperature!"

may enhance the sour feeling of acidic wines and retard the release of aromatic compounds. Thus, some people prefer their white wines at room temperature. The appropriate temperature is really the one you prefer, or are used to. I know of no other explanation for the nine-teenth-century appreciation of red Bordeaux at cool temperatures."

WINE TEARS

"Thus far, you've not talked about the formation of tears. How do they form, and does it have anything to do with the glycerol content of the wine?" asked Tony.

"The last part of your question is easier to answer than the first. Glycerol (glycerin) may have some minor influence, but is not crucial to the formation of tears, regardless of what you may have read. If you want to confirm this, add some glycerin to water and swirl. Although more viscous, no tears will form. However, water to which some Vodka has been added forms tears when the glass is swirled. Vodka is almost a pure alcohol solution. All that tears denote is the relative alcohol content of the wine. Another indicator of alcohol content is the intensity of the burning sensation it generates. Alcohol also produces much of the feeling of weight or 'body' in a wine.

"But, to answer the first part of your question on the origin of tears–tears develop after wine is swirled and forms a film on the inner surfaces of the glass. Because alcohol in the film evaporates rapidly, the surface tension of the film increases relative to the wine in the bowl. As the water molecules in the film pull together, they produce droplets at the rim. As the mass of the drops increase, they start to sag downward, producing the so-called 'arches.' Finally, the drops slide down, forming the 'tears.' When the drops reach the surface of the wine, some of the fluid is lost and the drops pull upward. Once formed, tears continue to develop–that is, as long as the film's surface tension draws up enough wine to counteract the effect of gravity."

WINE FLAVOR

"Well, back to tasting!" commented Craig. "While assessing taste and touch sensations, you also focus on flavor. Flavor refers to the joint sensations of taste and smell in the mouth. If you want to enhance your perception of flavor, draw air through the wine. This is what I call *aspiration*. For this, hold your head straight and draw air in through your clenched teeth. Air bubbling through the wine helps released compounds, similar to the practice of swirling wine in the glass. However, if you choose to do this exercise, practice in private, preferably in front of a mirror. If your head is too low, wine will dribble out between your teeth and down your chin. This is an excellent way to lose friends. However, holding your head back too far, to prevent the former situation, can result in wine going down the wrong tube. This will draw more attention to yourself than you probably want! Finally, if you are insufficiently discreet, people will think that you are slurping your wine. Don't hold me responsible if the *maître d'* throws you out of the restaurant!"

There were roars of laughter at Craig's graphic depiction of the various consequences of being overzealous in aspirating wine.

"Finally, we are at the finish." Craig snickered at his own pun. "With the 'finish,' we are concerned with taste and flavor sensations that linger in the mouth. Usually, the longer the sensation, the more highly rated the wine. The finish is the vinous equivalent of the fading sunset in summer. When resplendent, the finish can be more ravishing than the intense flavors produced in the mouth!" After Craig had let the image of his analogy sink in, he motioned that we were now free to taste the wine.

Everyone had enjoyed Craig's captivating presentation, and we were now anxious to try out his suggestions on the wines we'd brought. Even better was the realization that there was more to come. During the tasting, some of us continued to question Craig. Claude was still unclear on the difference between the bitterness and astringency of wine. Craig noted that Claude was not the only person to find separation of these perceptions difficult.

"The major problem is that both bitter and astringent sensations are produced by the same group of compounds, called tannins. They get their name from tanning leather. In the mouth, precipitated (tanned) proteins cause the dry, dust-in-the-mouth sensation we often experience in red wines. However, some small tannins also stimulate bitter-sensitive receptors on the tongue. Thus, both sensations occur together, especially in red wines, making clear distinction difficult."

"How do you succeed in swirling your wine so vigorously without spraying everyone in the vicinity?" Tony inquired. Craig had been swirling his wines with so much fervor that the wine repeatedly rode up and around the rim before descending, like a skate boarder on a ramp.

"The same way jugglers strut their stuff–practice! It's not hard. In my classes I demonstrate rotating the glass vigorously on a table or desk top. Once the students master this task, I tell them to continue rotating while they start to lift the glass off the surface. After some practice, students usually become proficient in vigorously swirling the wine, without anointing their compatriots."

Claude wondered, "What do you do at the dinner table? I assume you don't go through the elaborate ritual you've just described."

"No," exclaimed Craig. "I consume wine pretty much as most people, I suppose. What I do–and what most consumers don't–is consciously concentrate on the fragrance of the wine. However, I don't dissect the wine. The dinner table is a place to enjoy wine, the meal, and your company! The only exceptions are when friends come over to assess how several wines match with the food, or when a special bottle is being featured. In the latter situation, the wine is the central attraction, not a condiment to the cuisine."

After trying my luck at aspirating several wines, I decided that Craig was right–I needed practice at home.

Once people had finished sampling the various red and white wines we'd brought, Craig indicated that he wanted to try something new–something he hadn't tried in his courses, or with the Bacchus Society.

WINE FAULTS

"In my wine appreciation course, I talk about many aspects of wine, but rarely do I mention wine faults. Normally, one assumes that people don't need to be trained to recognize bad wines. However, people often reject sound wine, but accept faulty wine, especially when expensive. I suppose this is because they're afraid to suggest that a famous wine might be bad. In addition, detection of faults is similar to color vision. Some people are color-blind, while others have varying degrees of color blindness. Thus, wine that's revolting to one may be perfectly acceptable to another. Even when a person detects an off-odor, the subjective reaction to it can vary. I, for one, detect the fusel odor in ports, but don't consider it unacceptable like others. Also, it's not uncommon for people to incorrectly identify faults. Often those unaccustomed to wine misinterpret bone dry wines as being vinegary. Even those acquainted with wine often use the term 'corky' incorrectly. In addition, many faults are so unfamiliar that consumers can describe them only as 'not smelling right.' So tonight, we will try the faulty wines you brought, plus some that I will make faulty. If you want to clean your glasses, and bring your faulty wines, I'll dash out to the car to get my samples."

When Craig got back, he marked our seven faulty samples. He then requested that each of us pour a small sample in our glasses and order them in sequence. This done, he asked us to smell and taste each, noting if we thought the sample were faulty, and if so, what fault(s) it showed. When we were through, Craig collected the results. On glancing through the responses, he noted that it seemed that we'd tried different wines. For the same wine, we had given comments varying from "disgusting" to "no fault detected." There was also a scattering of named faults–including oxidized, vinegary, sour, sulfur, and corked–but with little consistency.

"It looks as if a bit of fault training would be useful! If you push your glasses back, I'll perform some vinous Black Magic. I'll turn good wine into bad. I wish I could do the reverse as easily. I'd be a millionaire by now."

We were intrigued as much by Craig's intent as by how he was going to achieve it.

Craig produced several 300 ml, screw-capped bottles of white wine from the box he'd just retrieved. Initially, he passed some around so that we'd remember the straw color of the unmodified wine. Subsequently, he opened a small vial containing a powdery white solid. He added it to one of the bottles, rescrewed the cap, and shook the sample. Craig laughed at our wide-eyed reaction to his shaking of the sample. He then passed the doctored bottle around with an unmodified sample. The treated wine had lost much of its color. He next poured some of the treated wine into a glass. Before passing it around Craig admonished us, "Don't take a strong whiff!"

When I received the glass and raised it to my nose, I understood Craig's warning. It certainly cleared my sinuses!

"Notice the potent burnt match odor that characterizes overly sulfured wines. What I added was potassium metabisulfite. It's the same ingredient that occurs in the Camden tablets people use in home winemaking. Not only does the sulfur dioxide released by metabisulfite protect wine against oxidation, it decolors the wine. As the wine ages, sulfur dioxide reacts with other compounds in the wine and becomes nonvolatile, or the gas escapes from the wine. In either case, the wine slowly regains its original color."

To another bottle, Craig added the contents from a second glass vial. This time, the concoction was a colorless liquid. When combined with the wine, it resulted in a slight but detectable darkening. On smelling, it was decidedly odoriferous.

"Here we have an instantaneously oxidized wine," remarked Craig. "What I've added is the major by-product of oxidation–acetaldehyde. A rough sherrylike odor should be evident. Note how readily the mild fruity fragrance of the wine is masked by the acetaldehyde. Note also the deeper color of the wine."

We were intrigued by the ease and rapidity with which such characteristics could be generated.

"I'd like to continue bewildering you with color and odor changes, but that's the extent of my instant sorcery." Bending down to reach for another sample, he said, "This sample was prepared by leaving the wine in the attic over the summer. Note the dark color, brown sediment, and strong baked flavor. If you are familiar with Madeira, the sample's resemblance to that wine style will be apparent.

"If you found my previous samples not entirely repulsive, I'm sure the following will be. Not that you were supposed to enjoy any of the previous samples, but some people do like the smell of sulfur, oxidation, and baked wine. Because I suspected that some of your samples might be corked, or somewhat vinegary, I have prepared some samples showing these classic faults. One contains a chemical called 2,4,6-TCA.[2] Even at trace amounts, it produces a moldy chlorophenol odor. My vinegary sample was produced by adding two teaspoons of vinegar to the wine a few weeks ago. The wine's sour odor is different

---

[2] 2,4,6-trichloroanisole

than the choking odor of pure vinegar. The additional off-odor is generated by ethyl acetate, the product of a reaction between alcohol and acetic acid (vinegar)."

"I'm glad we're not expected to taste *these* samples," I remarked. "Their odor is adequately informative!"

Craig chuckled and responded, "Now that you have tried my samples, let's go back to your samples and reassess their faults."

Tony had done remarkably well in identifying the faults. However, the majority of us—myself included—had mistaken most of the faults. We compared the faulty samples with Craig's samples. This was a new and fascinating experience for me. It was a great idea for the course I wanted to give when back in Brandon. During the discussion Phil asked, "Craig, you obviously have done considerable reading on faults. In your study, have you noticed anything on the origin of metallic tastes in wine?"

"Funny you should ask that!" responded Craig. "A friend showed me an article some time back arguing that the metallic sensation is an odor, not a taste."

"An odor?" asked Pat.

"I wondered about that," responded Craig. "But last week, I had a chance to verify the claim. I was having a sparkling wine at a party when, sure enough, there was a metallic taste. As I could hardly hold my nose in public while tasting the wine, I made a quick trip to the john. In the privacy provided, it became obvious that when I held my nose, the metallic taste did not appear, but developed only when I opened my nose. This does not explain the origin of the metallic sensation, but is a good example of how people can be fooled about tastes, odors, and flavors. Just think of the apparent sweetness of many fragrances. Where does their sweet 'taste' come from?

"There are a few faults I decided not to demonstrate. They were too vile! Regrettably, such odors occasionally do occur in wine. Examples are the sewer gas odor of hydrogen sulfide, or the rotten onion, sauerkraut, or sweat socks odors of different mercaptans. While instructive, I did not think you needed to smell these to recognize them. The names alone are sufficiently descriptive."

We all concurred! The few bad samples we had had were sufficient to illustrate the types of faults that can defile a wine.

"Am I to assume you are rejecting the wine, sir?"

At this point Craig nodded to Dave, and then turned to us. "Because you have been such good sports, and have played cooperative guinea pigs, I've one more sample for you. This time, however, it's a good bottle–a 1971 Gran Coronas Black Label from Torres. After what I've put you through, you deserve a reward."

And what a reward! I had read about this famous Spanish wine. It had been ranked above the most expensive and renowned Bordeaux wines in a tasting by French experts. It was a memorable conclusion to an outstanding evening!

Chapter 7

Thoughts on Wine Quality, Aging, and Fraud Detection

When I came in, Tony and Phil were grumbling about the blizzard that had clobbered us the day before. The town had been paralyzed for twelve hours, and many of the streets were still impassable. As long as you didn't have to venture out, the fairyland outside was a joy to behold. The trees were so heavily laden that their arched branches were planted in wintry drifts. Their spans were silhouetted against the azure heavens. Suzanne and I were glad we had brought our snowshoes. We had been out earlier in the day, relishing the warmth that had followed the storm. We had braved the drifts in the hope of picking up fresh cinnamon bread. Miraculously, Tiffany's Bakery was open.

Without our webbed footwear, I might not have made the meeting. The station wagon was still half-buried in the snow. Upon arrival at Dave's, I collapsed into the checkered couch next to the roaring fire. With the sun down, the damp had chilled me to the bone. The jolly crackling of the logs lifted my spirits as much as the flame warmed my body. After listening to the conversation for a few moments, I said, "You sound like a bunch of Canadians."

Deflecting his eyes from his hands, which were blistered from shoveling the driveway, Phil asked, "Why Canadian?"

"Weather is our national pastime. If there's nothing else to fill the verbal void, the conversation gravitates to the weather—past, present, or future. In Manitoba, we jokingly say there are ten months of winter . . . and two months of poor skating."

Tony responded, "Isn't it always so in Canada? We're always getting rotten weather from up north!"

"Whoa!" I exclaimed. "Have you got it wrong! Those blustery lows and frigid highs come from Siberia. You should be thankful we're there to buffer you from those arctic blasts! However, to change the subject, is everyone coming tonight?"

At that moment, Pat opened the front door and stomped in, trying to dislodge the white stuff solidified in the tread of his boots.

"As far as I know, everyone is coming," responded Dave. "At least, nobody has phoned to say they couldn't come. Even Klaus will be here. He was supposed to head back to Geneva, but the route is treacherous and access to the vineyards blocked."

While I rummaged through my backpack, Phil asked what I was looking for.

"A half-bottle of Inniskillin Icewine from Ontario," I mumbled. "Our present conditions seem appropriate to trying such a wine."

"It'll certainly be appropriate!" bellowed Pat from the hallway. "Is it the Canadian national drink?"

"Canadian wineries would love that, but as here, I'm afraid the national drink comes in a can, not a bottle."

Within a few minutes, Klaus arrived, followed shortly thereafter by Claude. Dave's wife, Mary, made a brief appearance with vegetables and dip and then vanished. Dave had as yet to convert her to the pleasures of wine. Although she took a little wine with meals, she had a reticence about drinking. Whether this was based on religious conviction, dislike of the taste, or something else, I never did find out. I was glad Suzanne enjoyed wine as much as I. Savoring wine by oneself precludes the pleasure of sharing the joyous experience.

It was Claude's turn to coordinate tonight's activities. Because we hadn't thought of anyone special to invite, he had arranged for three of us to do a short preparation of our own choosing. Pat had chosen the topic of quality, I had picked aging, and Claude had selected fraud detection.

CONCEPTS OF WINE QUALITY

"In my readings," Pat commenced, "there appear to be three principal ways of viewing wine quality: subjective, historic, and artistic. The concept everyone uses is the *subjective*. It's the 'I like it, I don't like it' sentiment that directs most wine purchases. This view is certainly reasonable and acceptable when based on personal

"It's a good wine, but not a great wine."

reaction to odors and tastes. Regrettably, it too often is governed by learned prejudices."

"Last time, Craig implied that there was considerable variation in people's sensory acuity," noted Claude. "Is there really that much variation in people's ability to taste? Or is it just that they lack the experience or attention to detail?"

"There probably is some psychological basis to people's enchantment or aversion to wine, but I know nothing on the subject. However, as far as measurable sensitivity to odors and tastes, the scientific literature is clear. There are even terms referring to people's inability to detect specific tastes (ageusia) and odors (anosmia).

These are equivalent to the types of color blindness. More fascinating, though, is the discovery that repeated exposure may improve the ability to smell some compounds!"

"That's neat!" declared Claude. "If we keep training, we may get better, not only because of experience, but also because our sensitivity increases."

"If that is so," cautioned Phil, "maybe we shouldn't have too many sessions on wine faults. We may end up super sensitive to off-odors, and find fault in all wines."

"A second concept of wine quality," continued Pat, "could be called the *historic*, traditional, or regional. It applies more to European than American wines. As you are aware, most European wineries produce wines from only one or a few local grape varieties by traditional procedures. This has led to the naming of their wines by geographic origin. The Appellation Control (AC) system maintains this situation, as noted by Peter several weeks ago. The ranking of the wine depends on the expression of the region's traditional flavor characteristics. These have evolved over decades or centuries. In contrast, many non-European wineries produce wines from a very wide range of grape varieties. Thus, wine quality is based on varietal aroma and flavor complexity, because clear regional characteristics have yet to develop. "

"But what about Californian appellations? Aren't there clear differences between Sonoma, Napa, and Monterey wines? I've read wine critics referring to regional Californian styles," noted Claude.

"Such styles may exist, but I've never noticed them. The last tasting at the Bacchus Society did nothing to change my opinion on the issue either. Often, I wonder if regional styles are nothing more than a figment created by critics needing something to say. In most cases, I feel the 'style' people detect comes more from the winemaker or grape variety than the region."

"I often wonder why so many people seem concerned about the views of others," said Tony. "Don't they have faith in their own abilities?"

"I think it's as Peter Phillips implied some weeks ago," noted Phil. "People want the best wine, but may have neither the time nor inclination to explore it for themselves. They depend on self-declared experts to direct their purchases. What these people don't

"It must be a coastal wine. . . . I can hear the ocean!"

realize is to what extent opinion is based on cultural precepts and experience. Personally, I only take note when critics comment on wines I've tried. Writers with whom I agree probably sense wine like myself."

"Another way to rate wine critics is whether they talk sense," commented Claude. "I remember reading an article by a supposed expert about three groups of Italian red wines. I compared the terms used by the writer for each group of wines. Hidden within the verbal diarrhea were the same descriptive terms. With the verbose subter-fuge removed, Chiantis, Barolos, and Valpolicellas all smelled of cherries and violets, with hints of tar. Regrettably, only readers familiar with these wines would know how ridiculous the comments were."

BREATHING

"While we're bashing inanity," said Phil, "I know of another good example of vinous nonsense. I once read a report noting the precise breathing times required for each of several related wines. Wine enjoyment was reduced to clock watching. From studying nineteenth-century cookbooks, I've the impression that the concept of 'breathing' had its origin in decanting. In the past, bottled wine often had considerably more sediment than today. Decanting had the benefit of both separating the wine from its sediment and getting rid of off-odors. These were lucidly described in the old literature as *bottle stink*."

"The view that wine improves in flavor," continued Phil, "rather than just losing its off-odors by breathing appears to have evolved early in this century. Presumably, the idea was concocted to explain the tradition of decanting. Currently, opening the bottle several hours before serving has supplanted decanting as the means of allowing the wine to breathe. I have seen no evidence that this works! In addition, tests conducted by myself and my students showed no benefit from decanting or early bottle opening. The only noticeable changes have been the slow development of an oxidative odor. This is quite different from the desirable release of aromatic compounds when swirling wine in a glass."

"That seems to correlate with what I've read about tasting very old wines," I mentioned. "Authors usually talk about a subtle bou-quet that fades within minutes. I suppose old wines, having little fragrance, quickly lose what they have. In contrast, younger wines possess more aroma and last longer after being opened."

"If I can get a word in edgewise," interjected Pat, "I'd like to complete my presentation. Then, you can have the floor all you want!"

With that prompting, we straightened up and became all ears.

"Now that I have your attention again," Pat said, chuckling to himself "the final concept of quality I want to mention is called *artistic*. It represents the ideals of balance and harmony—what Craig talked about last time. That is where no single sensory aspect dominates the wine. This implies that the wine is complex in flavor. Ideally, the flavor should also develop. Some authors call this 'opening up.' The analogy of an unfurling rose bud and the release of its perfume is often given. Remember the Vinho Tinto from Portugal I brought to our first meeting? Wine development also relates to Phil's comment on breathing. Allowing the wine to breathe may result in missing the flowering of the wine's fragrance. If the wine shows prolonged development, it means that the fragrance and flavor will last. This is often where great wines stand out. Inexpensive wines may be pleasant on pouring, but their fruitiness soon fades, leaving a nondescript wine. When a wine expresses all the attributes of greatness, it is truly a memorable experience!"

After a pause, Claude asked if there were any questions or comments. We looked at each other with that shrug-of-the-shoulder look, until Dave broke the silence. "We've been so impressed by Pat's crystal clear presentation that we're now speechless."

Pat peered out the corner of his eye as if to say "You're full of it!"

WINE AGING

"If we're done with quality," Pat announced, "its time for Ron to dazzle you with the vexatious enigma of wine aging."

"Is that what I'm going to do?" I questioned, eyebrows raised. "Well, we'll see! A few weeks ago, Peter Phillips mentioned several ways we might guesstimate wine aging potential. However, when I conducted a literature search on wine aging, I was astonished at the comparative lack of scientific work on the topic. In contrast, there is a deluge of advice in the popular press.

"Currently we know more about why most wines don't age well for long periods, than why a few do. Thus, what I have to say will not greatly aid you in selecting wine to age in your wine cellar. The information should, I hope, make you more appreciative of those few wines that do age well.

"I'm just going by the label that says this wine is best drunk when under three years old."

"Initially, it might be useful to mention why wines that age well are so highly regarded. We don't prize ten-year-old cans of peas, nor covet quarter-century-old steaks. Few grape varieties produce wines with long aging potential. Even here, production conditions greatly influence this property. Combined with human impatience, most wines that can age well never reach a venerable age. Thus, excellent old wines are rare, and correspondingly expensive. This gives such wines the appeal of an original work of art. For the wealthy, exclusivity makes venerable old wines almost irresistible. Whether right or wrong, owning a wine that others cannot obtain is alluring to many.

"The feature most commonly associated with aging is browning. In red wines, slight in-bottle oxidation slowly turns the pigments and tannins a reddish yellow/brown. In white wines, deepening of the color results from the oxidation of tannin subunits, or complexes that slowly form between sugars and amino acids. Deposition of the pigments as sediment or on the sides of the bottle occurs slowly, if at all, at cool temperatures."

"Do these changes drastically modify the taste of wine?" asked Klaus.

"Usually they are linked to a mellowing of flavor, associated with a reduction in the astringency of most young red wines. Their influence on white wine is less certain. If anything, aging may result in a subtle decline in sourness due to reactions between acids and other compounds in the wine.

"Although changes in color and taste can be important, aging most significantly alters the wine's fragrance. The first of these changes is beneficial–that is, the loss of the yeasty odor of young wines. Subsequently, white wines begin to lose the fruity character produced by certain esters formed during fermentation. Other esters develop during aging, but they don't contribute significantly to the evolution of an aged bouquet. Other fruity/floral compounds, technically called monoterpenes, tend to oxidize, changing their odor and becoming less volatile. Changes in the nature of other volatile compounds, found only in trace amounts, are probably involved. These compounds are thought to give premium grape varieties their characteristic aroma. Regrettably, clear evidence of their involvement in aging is only preliminary."

"Before you roar on, would you please let us laymen know what *volatile* means?" probed Tony.

"Sorry, Tony!" I responded. "Volatile refers to the ability of a compound to escape into the air at normal temperatures–the more volatile a compound, the greater its ability to reach your nose."

"Do the same events also occur during the aging of red wine?" asked Phil.

"Possibly," I replied, "but they seem less significant. Because red wines are fermented at higher temperatures, they contain lower levels of fruit-smelling esters. Thus, the importance of esters to red wine flavor seems minimal. Furthermore, few red wines seem to contain aromatic terpenes."

"If the aroma of red grapes does not come from esters and various terpenes, what is the chemical nature of red wine aromas?" asked Pat.

"In most cases, it's not known," I admitted, lifting my hands in defeat. "The pigments characteristic of red wines may be involved, but their importance in aging is unclear. In some cases, complex organic compounds, with even more complex names, are important. However, how aging influences them is unknown."

"Do researchers at least know the origin of the pleasant aged bouquet that develops in great wines?" pursued Pat.

"Not that I know of," I said. "In white wines, compounds with long-winded names form slowly during aging. Up to a point, they may contribute to a desirable aged bouquet. One such compound is 1,1,6-trimethyl-1,2-dihydronaphthalene, or TDN for short. However, as its concentration builds up, TDN donates a kerosenelike odor. More potentially pleasant is damascenone, if it accumulates sufficiently. Damascenone possesses a roselike fragrance. Also, bound (nonvolatile) forms of some aromatic compounds are known to be released during aging. However, the nature of the coveted 'cigarbox,' 'leather,' aged bouquet of some old red wines is still a mystery. In general, if the aromatic compounds that replace varietal aromas are esteemed, the wine is considered to age well."

"Books commonly recommend that wine should be stored under cool, dark conditions, away from excessive humidity and vibration," noted Tony. "Is there any evidence that such conditions are necessary? Wines stored in warehouses and bottle shops certainly don't receive such pampered treatment."

"The recommendation for cool temperatures–about 45° to 50°F (7° to 10°C)–reflects the temperature of unheated, underground cellars. This is the temperature range traditionally used for wine storage in Europe. Aging times are also generally based on storage at such temperatures. Whether you want to age at such cool temperatures depends on how patient, or how old, you are. If you intend to store young white wines for several years, holding at cool temperatures makes sense. In white wines, the fragrance of fruit esters is rapidly lost at the temperature most people keep their homes. Wine aging can be accelerated by exposure to high temperature (>100°F, 38°C), but the flavor is less subtle and desirable."

"What about darkness and humidity?" said Pat.

"From what is known, prolonged exposure to light is detrimental, especially for white wine in clear bottles. It's strange that the group of wines most adversely affected by light are generally the least protected. Light increases the rate of oxidation and may generate various sulfur off-odors. Sunlight is more damaging than artificial light because of the higher intensity and blue/ultraviolet content of solar radiation. Sitting in the sun rapidly increases the wine's temperature and thus volume. The pressure exerted can break the seal between the cork and the neck. This permits air and microbes to enter the wine, speeding oxidation and promoting microbial spoilage. Moderate humidity in the storage area retards drying of the cork, and limits wine seepage from the bottle. Excessive humidity, of course, favors the growth of molds that can soon make wine labels unreadable."

"Is there any truth to the contention that vibration disrupts or speeds wine aging?" asked Dave.

"To the best of my knowledge, no! In the past, moderate vibration might actually have been beneficial. It could have favored early wine clarification by promoting the settling of sediment. The regular production of crystal-clear bottled wines is a relatively modern development."

"In a book I read on Burgundy," said Phil, "the author complained about producers pasteurizing their wines. He implied that pasteurization stops proper aging because the wine is then no longer living. Have you read anything on the effect of pasteurization on the aging potential of wine?"

"I've read nothing specifically on wine pasteurization. However, my knowledge about the procedure–which refers to a wide variety of treatments–suggests that pasteurization should not affect aging. After all, aging is purely a physical and organic chemical process. The term *aging* is a poetic allusion for what is chemical maturation.

"Pasteurization is used to kill spoilage microorganism. Nevertheless, the conditions of the procedure are important. The lower the treatment temperature, the less likely the wine will suffer heat-induced flavor changes.

Changes in Wine Fragrance During Aging

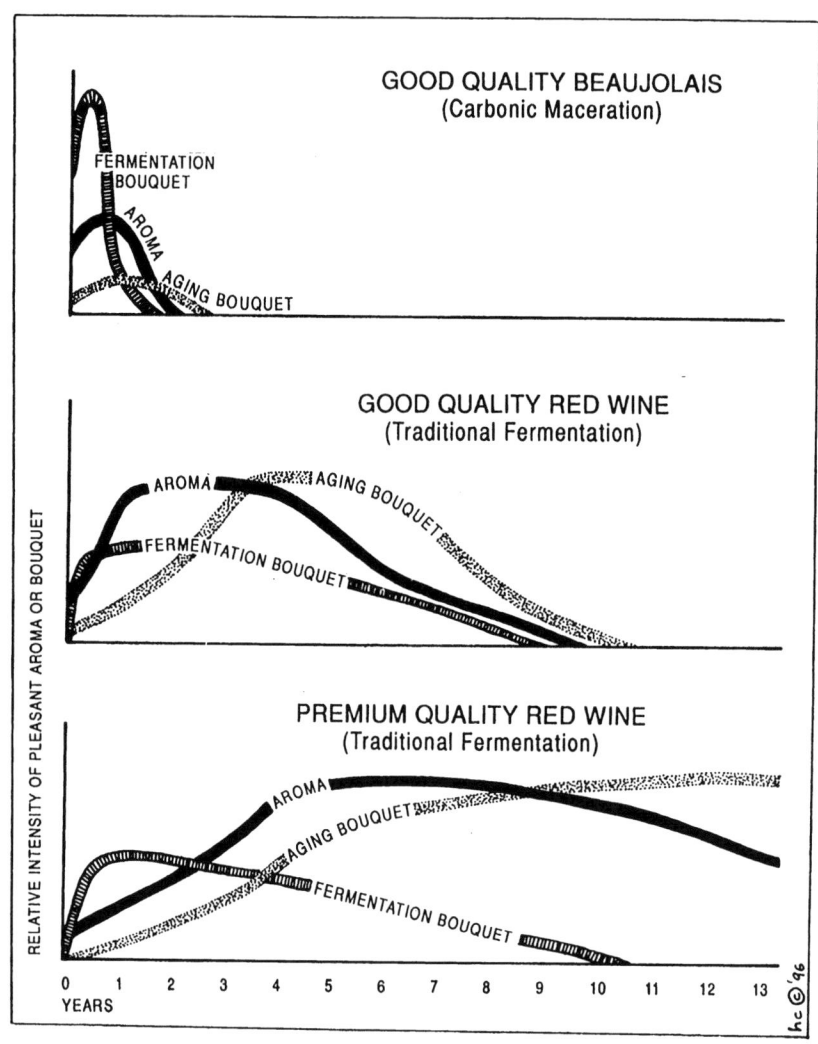

"Furthermore, wine is, in the strict biological sense, non-living. Although unromantic, wine is simply a savory solution of slowly reactive chemicals. If living processes do occur during aging, it's due to the action of spoilage yeasts and bacteria. This activity is definitely not wanted!"

"Now that you've mentioned chemistry," asserted Tony, "did you find anything on the effect of sulfur dioxide on wine aging?"

"If anything, it may slow aging by limiting some oxidative processes," I replied. "In addition, sulfur dioxide protects against spoilage. Nevertheless, in excessive amounts, it can be detrimental. Remember the color loss when Craig Goodwin added sulfur dioxide to his wine. Although the bleaching is reversible, sulfur dioxide may also indirectly cause permanent color loss. This results from retarding reactions between pigments and tannins."

"I know you said that your presentation wouldn't help us much in knowing which wines age well, but didn't you find anything useful in that regard?" prodded Pat.

"Possibly a few things," I pondered. "High tannin, alcohol, and sugar contents, as well as high acidity, favor long aging potential. Extended skin contact before or during fermentation–for white and red wines, respectively–also helps. This information translates into yellowish to golden white wines and almost black red wines, wines aged in oak, most fortified wines, and very sweet, acidic white wines. But these points you probably already know."

"The only way for consumers to know how a wine is aging," volunteered Klaus, "is to open a bottle every once in a while."

"And if you have only one bottle?" countered Pat.

At this, Klaus leaned forward, planting his elbows firmly on the table and said, "Then you're up the creek without a paddle!"

"Not exactly." Phil laughed. "One can still use the suggestions Peter Phillips gave us several weeks ago, such as cork length, grape variety, and so forth."

At this point I indicated that I'd finished my talk. Claude asked if there were any additional questions for me. With no queries forthcoming, Claude started to ruffle through his leather folder for the graphs he had intended to bring, but had forgotten.

WINE FRAUD

"Except for the criminal addition of toxic substances," began Claude, "consumer complaints seldom provoke governmental investigation. Most consumers, and probably wine judges, correctly assume that atypical flavors result from poor growing conditions, faulty corks, microbial spoilage, or some problem during shipping or storage. Even if wine is returned to the bottle shop, the most that usually happens is that the customer won't purchase the wine again. In most cases, fraud is detected through anonymous tips, unusual chemical purchases, or anomalies in winery records. At this point, chemical technicians must take over to establish the nature of the adulteration.

"It's important that the type of fraud be suspected, as wines can contain over five hundred compounds. Potential additives number in the thousands. In some cases, it's not the presence, but the atypical occurrence of a compound that denotes adulteration. For example, apple juice added to stretch grape must gives the wine an uncharacteristic level of chlorogenic acid."

"It seems then that fraud primarily involves contavening regulations," noted Klaus.

"Exactly," responded Claude. "Most fraudulent adulteration involves nontoxic compounds. This partially explains why consumer complaints rarely prompt governmental enquiries. Most wine drinkers don't detect, or misinterpret, changes in the wine's character. Even experienced wine judges can easily mistake the varietal, geographic, and vintage origin of wine. What chance, then, does the ordinary consumer have of detecting adulteration?"

"If that's the case, is wine adulteration more common than we suspect?" questioned Klaus.

"I doubt it. Most winemakers are very concerned about the quality of their product. They want to be proud of their wine, and this comes only if the wine's quality is based on using accepted procedures."

"But, if adulteration is hard to detect, how can one be certain that it doesn't occur?" Pat enquired.

"Faith," replied Claude. Detecting chemical doctoring is expensive and few labs are equipped to do the necessary tests. Investigators must know what they're looking for. The analytic equivalent of a fishing trip would be prohibitively expensive. However, current techniques can potentially establish the probable vintage, varietal origin, vineyard location, as well as the addition of nongrape juice, water, sugar, alcohol, and a host of artificial and nature-identical flavorants."

"*Artificial* I understand, but what on earth is *nature-identical?*" probed Tony.

"It's one of those expressions created by legal bureaucrats to describe natural compounds produced industrially."

"If they are identical to natural compounds, how can you tell them apart?" remarked Dave.

"When plants synthesize compounds, such as sugar, they incorporate radioactive carbon dioxide from the atmosphere. In contrast, similar compounds produced in the laboratory usually are made from fossil fuels. These sources contain only negligible traces of their original radioactivity, making modern and fossil-fuel-based compounds readily detectable by their different levels of radioactive carbon."

"But what if the compounds were obtained from other plants?" remarked Pat. "How could you detect if the alcohol in wine came from grape sugar or from added cane sugar?"

"It would be hard to identify the precise source of a suspect compound. However, to be 'foreign,' you need only establish that it could not have come from grapes. Because each living organism uses isotopes slightly differently, the compounds produced by them can be differentiated isotopically. Researchers can establish if a suspect compound could have been produced by grapes or during fermentation, from its isotopic ratios of carbon, hydrogen, nitrogen and sulfur."

"That's incredible," asserted Phil, shaking his head. "But how does one validate features such as grape variety, vineyard location, or dilution with water?"

"May I interject for a second?" asked Dave. "You've mentioned the term *isotope* several times. What exactly is an isotope?"

"Excuse me!" apologized Claude. "I am so familiar with the term, I forgot it's not part of everyone's vocabulary. Most chemical elements exist in one or more forms that differ only by weight. Most of you have heard of 'heavy' water used in nuclear reactors. It gets its name from the heavier form (isotope) of hydrogen found in the water molecule. Several heavy isotopes are unstable and break down gradually. Such isotopes are called radioactive. Is that sufficient?"

Dave nodded his approval and appreciation, so Claude returned to Phil's question.

"The ability to confirm varietal makeup depends on the grape cultivar. The aroma of some grape varieties is sufficiently distinctive to permit a tentative identification of a wine's varietal origin. However, because a wine's chemical makeup changes during aging, comparison with authenticated examples of the wine is necessary.

"Even the vineyard location can potentially be validated—by assessing the isotopic ratio of the water in the wine. Water from different regions display unique hydrogen and oxygen isotopic ratios. In addition, plants tend to selectively use the lighter molecules of water. Thus, the isotopic ratio of water in grapes (and wine) differs from its source, the soil. Therefore, even dilution of the juice or wine with water from the vineyard can be detected!"

"With all this forensic sleuthing," noted Pat, "you may be correct in suggesting that wine adulteration is not common."

"I have listened with particular interest to your talk," noted Dave. "Like everyone else, I'm impressed with the power of modern analytical skills. However, except for the witless addition of toxic materials like antifreeze or methanol, what difference does it make if a producer adds a nature-identical flavorant to his wine—especially if it improves the wine's flavor?"

"That's a good question!" Claude replied. "It primarily comes down to an issue of public confidence in the wine industry. Most winemakers want their wines to be considered a natural wholesome beverage, not a chemical concoction. It's also a reaction to the fear many consumers have of chemical additives. This applies even to nature-identical compounds. It is curious how long-established fermentations like bread, cheese, and winemaking are considered natural, but the equal microbial production of vitamins, flavorants, and drugs is viewed as unnatural.

"Another aspect of the issue relates to the profit advantage that may be gained by adulteration. It puts honest producers at a distinct disadvantage. In addition, expensive wines are usually touted for their unique site-derived identity. Chemical doctoring would be viewed as invalidating its genuineness. Permitted modification is politely called *amelioration*, while modification designated as illegal is officially termed *adulteration*. What is and is not permitted varies from country to country, or even from region to region."

"So it's chiefly consumer image that drives the need to assure authenticity?" commented Tony.

"At least that's my view," responded Claude.

After a few moments of silence, Dave remarked. "Do we have any more questions for Claude? If not, we have an icewine from Ontario provided by Ron. I don't have a comparable German *eiswein*, but I do have a bottle of Joseph Phelps Late-Harvest Riesling."

It was great to relax with lovely wines after such a long meeting. We finished off with persimmon on clotted cream, topped with homemade oatmeal cookies, another of Dave's favorite treats.

Chapter 8

Visit from Dr. Nicholson:
Wine and Health

One evening will always stick out in my mind. Shortly after our regular walk to the Pyramid Mall, Suzanne and I decided to hit the hay early and do some reading before drifting off to the land of nod. I folded back the blankets, puffed up the pillows, propped them against the east wall, got comfortable, and began to peruse the latest issue of *Decanter* magazine. Suzanne was in the bathroom getting washed. I was partway through an article on Barolo when suddenly I heard a *pa-ouff*! My head whipped around to see if the noise had come from Suzanne's direction. At the same instant, Suzanne was looking at me, equally wondering where such a bombast had come from. Within a split second, another more extended ripping and hissing was heard. Suddenly I sank four inches. Impulsively I blurted out, "Now that's a let down!" Suzanne began to laugh so hard at my sudden descent to the floor; she could hardly see enough through her tears to find out what had happened. After flinging off the bedsheets, I confirmed our suspicion. A cell in the air mattress had ruptured, rending it asunder. At the same time the phone rang. It was hard to establish enough composure to answer. Nevertheless, I controlled myself long enough to get Phil's message about our next meeting. Then I had to relate why we were in such a hilarious state. Suzanne was still chuckling in the background.

It was a bit of a shock when I met Dr. Nicholson the subsequent Sunday afternoon. Phil had not advised us that our guest speaker was a woman. Instead of the eccentric, long-haired, bohemian, here was a handsome, bronzed lady in her early forties. She looked more like a model than an MD. She was impeccably, but conservatively, dressed in a navy blue pin-striped Edwardian jacket, with charcoal grey pants. She must have sensed that we felt out-of-place in our sweatshirts and jeans, as she commented that she was expected at the residence of the university president shortly after our meeting. There would be insufficient time for her to change before going to the dinner party. She was being recognized, along with several others, for excellence in university teaching.

Phil introduced Dr. Nicholson as an allergist affiliated with the Cornell Medical School in New York City. She occasionally came to Ithaca for consulting and to visit her brother, a professor in the Physics Department. Phil had met her through his interest in food allergies. Dr. Nicholson was a specialist in food and beverage allergies and related immune problems.

WINE-INDUCED HEADACHES

"Because I assume you're more interested in the positive health benefits of wine consumption than its abuse," she said, "I shall concentrate on the former. Nevertheless, I should mention one of the most common problems linked with wine consumption–headaches; and here I do not mean a hangover. The headaches my patients see me about are those connected with the consumption of moderate to minute amounts of wine. Although the causes may be diverse, the situation is becoming somewhat clearer as more practitioners take these complaints seriously. Classifying headaches by their symptomology will help us discover their diverse causes. Knowing the precise origin of each type of headache should assist in the development of effective treatments.

"Of wine-induced headaches, the most well-known is the migraine. For a time, people thought it might be provoked by the presence of biogenic amines, such as histamine and tyramine. However, they do not occur in wine in sufficient amounts to cause a migraine. In addition, double-blind experiments do not support the hypothesis that amines are the major factor in wine-induced migraines."

Tony raised his hand slightly to attract Dr. Nicholson's attention. "What is a double-blind study?"

"To put it simply, a double-blind study requires that neither the participants nor the administrators know which test substance is being given to whom. Thus, results cannot be biased by knowledge of what was being given. In the particular study I mentioned, participants received wine samples with or without added histamine. At the same time, the participants were given either antihistamine or a *placebo*. Does that clarify the nature of the test?"

As Tony indicated his agreement, Dr. Nicholson turned to the rest of us and encouraged us to ask questions. She admitted that she was more familiar with discussing issues like this with student physicians and colleagues than wine groups. Thus, she might inadvertently use technical terms unknown to us.

"Although the amine content appears insufficient to cause migraines, the tannins and other phenolic compounds in wine may trigger migraine induction. The body destroys most phenolic compounds that enter the blood with the enzyme PST (phenolsulfotransferase). However, people sensitive to food-induced migraines often have low levels of platelet-bound PST. Phenols induce platelets to release 5-HT (5-hydroxytryptamine). The same compound, commonly called serotonin, acts as an important neurotransmitter between nerve cells in the brain. In blood, 5-HT causes platelet clumping that can block small vessels. 5-HT also activates the constriction of small blood vessels in the brain, causing pain and other symptoms associated with a migraine. People also may show cyclical patterns in platelet sensitivity to 5-HT release. This may explain why consuming red wine is not consistently linked to migraine induction. Small phenolic components in wine also prolong the action of potent hormones and nerve transmitters such as histamine, serotonin, dopamine, adrenalin and noradrenaline. These can affect headache severity and other allergic reactions."

"If I understood you correctly," I commented, "the primary phenols causing problems are tannin subunits. What about the large polymers that form during wine aging?"

"Large tannin polymers, unlike their subunits, do not enter the blood. This may explain why aged red wines tend to be less associated with headaches than young red wines. A classic example is the ease with which the youngest of all red wines, Beaujolais nouveau, produces headaches.

"Researchers recognize at least two other wine-related headaches," she continued. "The *redhead* develops in the morning, about a hour after waking. It follows the consumption of even modest amounts of some red wines. It develops as a severe headache that, on reclining, causes nausea. It usually dissipates by noon. A similar phenomenon has been reported with some Californian Chablis or mixtures of white wine taken alone, or with coffee or chocolates. Its chemical cause is unknown.

"Another recognized headache syndrome is called the *red wine headache*. It may develop within minutes, its intensity often being dose-related. The headache reaches its first peak within about two hours, tends to fade, but returns some eight hours later in a more intense form. The headache seems related to the release of prostaglandins, important chemicals involved in constricting blood vessels. Thus, the development of such headaches is often prevented by the prior consumption of prostaglandin synthesis inhibitors. I have found that taking ASA (acetylsalicylic acid), a prostaglandin synthesis inhibitor, prevents my getting a headache at wine tastings."

"That's great for those who can take ASA," noted Dave. "However, what about people like myself who are sensitive to ASA?"

"You could try an enteric-coated ASA to avoid the stomach burning, if that's the problem. Otherwise, you could try one of the other prostaglandin synthesis inhibitors, such as Advil® (ibuprofen) and Tylenol® (acetaminophen). They seem to be as effective as ASA.

"Taking wine only with meals is a long-known precaution against wine-induced problems. Food in the stomach delays the movement of alcohol into the intestinal tract, where some 80 percent of the alcohol is absorbed. By slowing the rate of alcohol uptake, its absorption matches more closely the body's ability to metabolize alcohol. Correspondingly, the peak alcohol content reached in the blood is reduced. This also would delay the uptake and maximum level of phenols in the blood."

"I've read that the carbon dioxide content of sparkling wines speeds the uptake of alcohol," noted Pat. "Is that true?"

Pat had been quite taken with Dr. Nicholson. He had nudged me before the talk, commenting that were she his physician he would go to the doctor more often.

"I've heard that assertion myself," noted Dr. Nicholson. "If carbon dioxide relaxes the sphincter muscles that regulate movement from the stomach into the intestine, it would speed alcohol uptake, hastening flow into the intestines. However, I seem to re-member that this scenario has been refuted. The 'potency' of spar-kling wine may only reflect its being taken on an empty stomach.

EFFECT ON FOOD DIGESTION

"While on the topic of consuming wine with food, wine en-hances the release of the hormone gastrin. This promotes the pro-duction of gastric juice in the stomach. In addition to aiding food digestion, gastric juice inactivates enzymes involved in ulceration. Thus, although not an ulcer medication, moderate wine use can limit the development of stomach ulcers. In contrast, distilled alco-holic beverages depress the production of gastric juice and can cause stomach cramps.

"Several studies have shown that taking wine with meals im-proves the appetite of anorexic patients, reduces food consumption by the overweight, enhances self-esteem in the elderly, and taken just before bedtime, promotes sleep. Gastrin release may explain the first of these factors, but the nature of the other phenomena is unknown."

WINE AND ARTERIOSCLEROSIS

"However, the most well-publicized and scientifically substan-tiated benefit of wine consumption is increased life span. Recent studies show that those who consume wine moderately live, on average, two and a half years longer than teetotallers, and consider-ably longer than heavy drinkers. There is also strong correlation between wine consumption and reduced incidence of death due to coronary disease. What is especially encouraging about these studies is that the underlying causes for the statistics are now largely understood. However, before describing how wine and alcohol have these benefi-cial effects, I want first to outline the origin of arteriosclerosis, the major cause of coronary disease.

"Arteriosclerosis results from chronic injury to the arteries. Al-though associated with various factors, they all seem to favor the oxidation of certain cholesterol-containing blood proteins, the low density lipoproteins (LDLs). Oxidized LDLs directly damage the artery wall as well as promote the adherence and migration of white blood cells into the wall. Some of these accumulate LDL molecules. Slowly, smooth muscle cells accumulate in the artery wall and develop their own vasculature. Subsequently, these cellular accumulations become fibrous, inelastic, and protrude into the artery, restricting blood flow. This sets the stage for platelet aggregation, clot formation, and the blockage that can precipitate a heart attack or stroke. However, if the causal irritation (smoking, high blood pressure, high dietary sources of cholesterol, or certain bacterial infections) are eliminated, arteriosclerosis may be reversed. Part of the reversal relates to elevated levels of high density lipoproteins (HDLs). They appear to remove the cholesterol that has accumulated in the arteries. These proteins are the so-called 'good' (HDL) blood cholesterol, in contrast to the 'bad' (LDL) blood cholesterol. Moderate wine consumption, notably red wines, both increases the concentration of HDLs and decreases the level of LDLs in the blood."

WINE'S ANTIOXIDANT EFFECTS

"In addition to the beneficial effects of alcohol on LDL/HDL concentration, wine phenols suppress LDL oxidation. The antioxidant action of wine phenols, notably the smaller tannin subunits, also appears to protect cell structures from oxidation. The latter may slow the transformation of healthy cells into cancerous cells."

"I must comment," said Dave, "that I appreciate your lucid explanation of a complex subject. However, what does moderate consumption really mean?"

"It varies somewhat from study to study," said Dr. Nicholson, "but would equate to about 250 to 300 ml per day (about a third of a standard wine bottle). Various researchers measure intake in terms of alcohol content, while others calculate it in terms of the number of drinks."

"Have you read anything about resveratrol?" I questioned. "It's a phenolic compound produced by grapes in response to fungal attack. It is suspected to be an important antioxidant."

"What I've read suggests that it is more active than antioxidants such as vitamin E and ascorbic acid. However, it is both less com-

"Sidney's doctor has cut him down to one glass of wine at dinner."

mon and less active than flavonols such as quercetin derived from grapes."

At this, Tony began to wave his hands in the air. "Whoa! What are flavonols? Remember, we're not chemists."

"Excuse me," said Dr. Nicholson. "I slipped up there, didn't I? Most wine flavonols are tannin subunits. Quercetin is a related wine phenol. I don't know if that helps. I could give you the chemical formula, but I doubt that's what you want."

"No!" stated Tony, waving his hands. "That they are tanninlike molecules is sufficient. Thanks."

At this point, Klaus spoke up. "Of some interest may be the fact that the quercetin content varies widely in grapes and wine. The content depends partially on the duration of skin contact and the type of fining. For example, fining with PVPP[1], a fairly new clarifying agent, markedly reduces the quercetin content."

"That's regrettable," remarked Dr. Nicholson, "especially since quercetin is both one of the best antioxidant and anticancer compounds found in the human diet. Thankfully, most people obtain quercetin from fruits and vegetables."

WINE'S ANTIMICROBIAL EFFECTS

"Because phenols and tannins react relatively nonspecifically with proteins, would this explain wine's antimicrobial effects?" I enquired.

"Partially," responded Dr. Nicholson. "The actual antimicrobial effects of wine are still poorly understood. The effect of alcohol in wine is relatively minor, due to its low concentration. Currently, the most antimicrobial compounds that have been isolated are, as you suggest, phenols and tannins. Phenolic compounds such as *p*-coumaric acid are particularly active against gram-positive bacteria like *Staphylococcus* and *Streptococcus*, while other phenols are active against gram-negative bacteria like *Shigella*, *Proteus*, and *Vibrio*. The latter cause serious types of diarrhea and dysentery. Thus, it is not without reason that Roman armies added wine to the drinking water to avert intestinal disease. Wine is also effective against viruses, such as the poliovirus. Tannin polymers actively neutralize viruses. In this regard, it's interesting to note that consuming wine reduces the likelihood of catching a cold. Maybe we should gargle with port or Chardonnay instead of mouth wash," she said, chuckling.

WINE, LEAD, AND GOUT

"Dr. Nicholson, in the 1800s, there were many reports linking wine consumption, especially port, with gout. One seldom hears of this today. Was this association erroneous?" Dave asked.

"Gout is caused by the localized crystallization of uric acid and its inflammation of the joints. This results from lactic acid, formed

---

[1]polyvinylpolypyrrolidone

via alcohol, reducing the excretion of uric acid in the kidneys. Alcohol also raises blood uric acid content by promoting purine breakdown, a prime source of uric acid. Wine is less likely to induce gout than beer, because of its lower purine content. Nevertheless, medical historians suspect that lead-induced kidney damage is the primary cause of the gout epidemic during the nineteenth century. Samples of port from the 1800s often show high lead contents. Pickup from stills, used in the making of fortifying brandy, may have been the source of the lead contamination. In addition, pewter and lead-glazed drinking cups, and storage of wine in lead crystal decanters, can add significantly to the lead content of wine."

"In other meetings of our group, we have mentioned the pros and cons of sulfur dioxide addition to wine. What is your opinion on this subject?" asked Tony.

"I cannot comment on its use in wine production, but for most people, the sulfur dioxide content of wine is medically insignificant. However, for some asthmatics, sulfur dioxide may induce bronchial constriction. Sulfite, absorbed via the digestive tract into the blood, can constrict the bronchial passages. Even more at risk are those few individuals afflicted with a rare genetic disease, called sulfituria. They are unable to produce active sulfite oxidase, an enzyme that converts sulfite to sulfate. These individuals must live on a very restricted diet low in sulfur containing proteins. It is estimated that normal synthesis of sulfate from food is about 2.4 grams per day. The sulfites in wine contribute only marginally to this amount."

WINE AND CANCER

"You mentioned earlier that some tannins protect you against cancer. However, I have read that some tannins are mutagenic. Are not all mutagens carcinogenic?" inquired Phil.

"Surprisingly, no," commented Dr. Nicholson. "For example, the phenol quercetin causes mutations in laboratory tissue culture, but actively inhibits cancer production in whole-animal studies. This apparent anomaly may result from differences in the concentration of quercetin used, and the low level of metal ions and free oxygen found in the body. In addition, phenols (tannins) detoxify the small quantities of nitrites normally found in food, but convert the high concentrations of nitrite (a preservative found in smoked and pickled foods) into diazophenols. The latter can induce oral and stomach cancers.

"Moderate wine consumption is not linked with an increased risk of cancer, with the possible exception of a slight increase in the incidence of breast cancer. However, high rates of wine consumption have been linked with an increase in mouth and throat cancers."

WINE AND MEDICATIONS

"I know doctors tell you not to consume alcohol with certain medications," remarked Dave. "Are there any simple rules that can guide us concerning this issue?"

"Most of the literature relating to alcohol-drug interactions have been derived from alcoholics or binge drinkers, not those using wine sensibly. However, if you were using tricyclic antidepressants, you should be aware that small amounts of alcohol can cause loss of muscle control in some individuals. In addition, because of the frequent use of MAO[2] inhibitors in the control of hypertension, these patients should know that some red wines can reduce the effectiveness of the medication. As well, long-term use of acetaminophen can enhance alcohol-induced kidney damage. In general, though, it's best to communicate with your doctor, especially concerning any apparent abnormalities observed with medication you think might be linked with wine consumption."

WINE'S NUTRITIVE VALUE

"During your presentation, you have not mentioned wine's food value. Would you comment on this issue?" asked Phil. Although already knowledgeable on the subject, Phil wondered if Dr. Nicholson was familiar with any material of which he was unaware.

"I can't say that wine is especially nutritious, but it does contain valuable quantities of several vitamins, notably riboflavin, niacin, pyri-

---

[2]monoamine oxidase

doxine, and folate. Wine also contains several minerals in readily available forms, especially iron in the ferrous state. The low sodium/ high potassium content of wine makes it one of the more flavorful beverages for people on a low sodium diet. Wine also can be an effective source of potassium for individuals using diuretics. Furthermore, alcohol is a readily available source of caloric energy. Although not significant in our modern world, wine used to supply a considerable proportion of the caloric intake of peasant farmers.

"Wine also has several beneficial but indirect food influences. I have already noted the value of wine in stimulating the production of gastric juices and fostering a healthy appetite. Wine is also well known to promote relaxed eating, something increasingly valuable in our overly compulsive society. For example, the presence of γ-butyrolactones in wine may be even more important than alcohol in reducing stress."

FETAL ALCOHOL SYNDROME

"There is one topic I have not discussed, nor have you asked about," she said. "That is the fetal alcohol syndrome (FAS). If your wives had been here, I'm sure FAS would have been one of their first questions. In case they ask you about it, I'll give you some information on the issue.

"FAS refers to a set of symptoms including suppressed growth, mild mental retardation, and facial abnormalities. It is most often found in the children of alcoholic mothers. At the outset, I want to note that there is little or no evidence that moderate wine use during pregnancy is likely to cause FAS. The real cause of the detrimental effects observed in FAS is unknown. The problem is complicated as alcoholic mothers tend also to be heavy smokers, use illicit drugs, consume large amounts of coffee, and/or have poor nutrition. Although there is no *absolutely* safe level of alcohol consumption, or almost any food for that matter, there is no reason to suspect that taking a glass of wine with meals should harm the developing fetus. Equally, though, there should be no long-term harm if the mother abstains from drinking wine during pregnancy. Finally, it's useful to realize that the natural bacterial flora of the intestine produces alcohol during food digestion. Thus, alcohol is a natural component in our diet, even if we never drink a beverage containing alcohol."

"Your last comment is especially intriguing," commented Pat. "How is it that we don't become inebriated on the alcohol produced by our intestinal flora?"

"Admittedly, the production is small and discontinuous. Also, blood carrying the alcohol from the intestinal tract is routed first through the liver. The liver is the site where more than 90 percent of the alcohol is removed and metabolized. It is converted to acetaldehyde and then acetate. From that point, the breakdown products may be lost via the lungs or used in body metabolism."

"Do we all have bacteria in our intestines?" squirmed Dave.

Dr. Nicholson laughed. "Yes! So many that the number boggles the mind. Up to 30 percent of the dry weight of feces consists of bacterial cells. Don't be alarmed!" she said, raising her hand. "Instead, you should be pleased. Bacteria help digest your food, produce several vitamins, and assist in protecting you from several intestinal infections."

"I'll accept your word for it, but I'm still not inspired by the thought of a bellyful of bacteria," concluded Dave.

"I'd enjoy staying to answer more questions, but I see by the clock on the wall that I must leave to meet my brother. If there are a few more quick questions, I will try and answer them, but then I must depart."

"I have one quick question," commented Pat. "With all the evidence on the benefits of moderate wine consumption, are we likely to have the family doctor recommending wine in the near future?"

"Probably not in my lifetime." Dr. Nicholson smiled. "It's risky for doctors to recommend much these days. Even if most people consumed wine reasonably, there would always be those who would be unable to regulate their intake. People have enough trouble even eating sensibly. American culture has yet to reject the notion that sociability and maturity are associated with one's ability to hold liquor. In contrast, southern Europeans have developed a rational accommodation with alcohol in their diet. There, wine is viewed as the beverage to have with meals, not as a 'drink.' Children are taught by parental example to view wine as a food beverage. Drunkenness is an object of scorn and family shame, not viewed with furtive envy."

With that fitting conclusion, Phil expressed our sincere appreciation to Dr. Nicholson for having taken her time to come and clarify the issue of wine and health. At this point, Dave brought out a present–a bottle of Lindemans Pyrus from Coonawarra in Southern Australia. Phil had told Dave of Dr. Nicholson's preference for this wine. When she realized what it was, she gave Phil a long hard look. "Now I know why my brother was asking what my favorite wine was. Thank you very much!"

After Dr. Nicholson bid her farewell, we continued to talk for some time about what she had discussed. We all thought that this was one of our best, or at least our most informative, meeting to date. Our knowledge of wine was expanding by leaps and bounds. I felt very lucky to have made contact with such a group. I was also getting anxious to impart my newfound knowledge to others.

"After such a fascinating talk, it seems an anticlimax to mention our next meeting," noted Phil. "Nevertheless, we have to make some decision on what it should be."

"Thus far, we have not discussed wine-making techniques other than table wines. Why not have a meeting about sparkling wine making?" I suggested.

"Who should be invited, or where should we go?" questioned Claude.

"Why not go the Gold Seal Winery in Hammondsport," said Pat. "They certainly make a lot of sparkling wine. They may also be willing to discuss sherry and port production."

"I like the idea. That means that we'll have nothing to prepare and have free samples." Dave chuckled.

"Phil, can we count on you to arrange things with Gold Seal? I suspect you've better contacts with them than the rest of us. We'll probably get a higher ranking reception through you," asserted Dave.

"OK! OK!" said Phil. "I'll contact you when everything is set up."

Chapter 9

Visit to Gold Seal Winery

One never seems to remember how a conversation drifts in a certain direction. On our way to the winery, Suzanne and I got into a conversation about our relatives. Both families have had members opposed to–or at least fearful of–consuming alcoholic beverages. On my side have been Quaker ministers and many stalwart Presbyterians. What penitence would they have undergone to avoid having an offspring infatuated with the fermented fruit of the vine? Thankfully, my Anglican persuasion is more liberal, and my inclination not considered deviant behavior. Suzanne's side of the family generally likes wine, except for her mother. Mrs. Ouellet is concerned that we both may become alcoholics. We chuckle at her concern as we only take wine with meals. Wine makes us so sleepy we'd fall asleep before consuming enough to become tipsy.

We were about to lament the sorry state of those unable to regulate their alcohol intake, when we saw the diminutive village of Hammondsport appearing around a bend in the road. It was time to regard the directions Phil had given us for locating the winery.[1] From the size of the village, it was clear that it would take some doing to get lost.

Possibly because I grew up in the center of an industrial city, I have always yearned to live in a small town. Hammondsport fits my image of a warm, motherly hamlet where people take time to be friendly. Even its nestled location at the foot of Keuka Lake seemed to engender a sense of being at home.

On a Saturday morning in early spring, it seemed that the crocuses and snowdrops were the most active individuals in town. The

---

[1]The operations of the original Gold Seal Winery, now part of the Canandaigua Wine Company, have been moved to the Taylor Winery location in Hammondsport.

low sun angle gave the collection of mid-Victorian homes, pastoral farm buildings, and turn-of-the-century business section a charming aura I now equate with Upstate New York.

As we drove into the parking lot, I noticed Phil leaning with his back on the hood of his Mustang. He was taking the opportunity to enjoy the vernal warmth and obliterate the winter pallor. Tony and Claude had gone into the winery shop to do some preliminary enophilic surveying.

When everyone had arrived, Phil showed us into the office where John Winkle was waiting. John was a University of California (Davis) graduate with an intriguing and varied career. At various times, he had been assistant or chief winemaker at several Californian estates, private enology consultant, successful university lecturer, and now winery administrator. When people have such an extensive and impressive *curriculum vitae*, I tend to expect them to be white-haired. Not John. His boyish looks and casual attire made him look more like a lab technician than production vice president. I never fail to be amazed at the drive and ability of some of my fellow beings. I suspect "tortoises" like myself are always in awe of the "hares." Even more surprising was his lack of affectation. His warm, genuine smile and calm personality put us at ease.

After the introductions, John took us on a quick tour of the facilities. It was good to have a refresher on winery operation before getting into some of the specifics of port, sherry, and sparkling wine production.

PORTS: AMERICAN AND PORTUGUESE

"Before discussing how we produce ports," John began, "I want you to taste some of our products, and compare them with those from Portugal."

John motioned to the tasting room attendants to bring the glasses of port. Each of us was supplied with four ISO glasses containing deep- to brick-red wine. John asked us to sample the first two and then the last two. When we had finished the enjoyable task, John requested that we now compare the two groups.

Dave was the first to comment on the obvious color differences between the wines. The first two were deep red, with no brownish hues. In contrast, the second set possessed a brickish color.

"This indicates one of the major differences in how most ports are produced here in the United States versus Portugal," commented John. "In North America, the clarified, partially fermented fortified wine is exposed to heating (110°F, 45°C) for about a week, possibly associated with aeration. The postfermentation treatment, called *processing*, speeds aging and permits the wine to be marketed earlier than the two or more years required in Portugal."

"Do the different processing techniques explain the differences between the two groups of ports?" I inquired.

"Primarily," commented John. "Although the use of different grape varieties is influential, processing gives our wines their distinctive *rancio* flavor."

"Isn't that considered a fault?" commented Pat.

"Obviously we don't think so, nor do our customers. What some people consider a fault in one type of wine is coveted and traditional in another. For example, Portuguese ports are expected to show a fusel note, and sherries exhibit an oxidized character."

"Where does the obvious sweetness in these wines come from? Do you add sugar?" questioned Claude.

"You may not have grasped the significance of my statement that ports are made from partially fermented grape juice. That is, the fermentation of the juice is stopped halfway through by the addition of high-proof spirits. The added alcohol intoxicates the yeasts and fermentation ceases. This leaves the wine with about half the original sugar content of the juice. The residual sugar provides the wine's natural sweetness. Occasionally, though, the juice may be fermented to dryness to increase pigment and tannin extraction. In this case, both unfermented grape juice and spirits are added to supply the sweetness and desired alcohol content."

"This seems to imply that there are only red ports. However, I remember seeing white ports, in Portugal," interjected Dave.

"There are white ports, as you say, but we don't produce any here. Right or wrong, they remind me of sweet sherries," responded John.

"Isn't there a whole range of ports found in Portugal, not produced here?" I probed.

"Indeed! With few exceptions, American ports are fairly similar in character, but don't say I said that. The main differences reflect blending to establish brand identity. In Portugal, several distinctive styles have evolved over the years. The style I presented in comparison to our port was the most commonly produced–*ruby port*. In this style, producers age the wine in large casks for two to three years. Individual brands differ in the formulation of their base wines. Ruby ports are a blend of wines coming from several vintages and different vineyards. *Tawny ports*–at least the better versions–are blends of ruby port aged in oak for ten or more years. During aging, much of the red pigmentation precipitates or is oxidized to a brickish color; therefore, the origin of the tawny appearance and the name for the style. Inexpensive tawny ports are usually blends of red and white ports.

"*Vintage ports* are produced only in especially good vintage years and are bottled within two years. For this style, aging is largely in-bottle, and occurs in the absence of oxygen. Because of the minimal oxidation and clarification, the wine retains its red color for decades, throws considerable sediment in the bottle, and has a different bouquet than ruby or tawny ports. *Late-bottled vintage ports* are blends from a single vintage, aged similarly to ruby ports, but bottled within five years of the vintage. Sound complex? . . . It is!

"Initially, ports were primarily exported to England, where an increasingly sophisticated upper class supported the production of a range of port styles. In fact, most of the firms are, or at least were, British. In contrast, the American market has never developed sufficiently to support an equivalent diversification. California has a few wineries specializing in a range of ports and dessert wines, but none occur here in the East."

SHERRIES: AMERICAN AND SPANISH

Because we seemed to have exhausted ourselves of questions concerning port, John shifted his attention to sherry.

"Sherry, like port, evolved over several centuries. Initially, sherry was a light table wine produced in the southeastern region of Spain, around Jerez de la Frontera. Over the years, sherry became a fortified wine that has diverged into several distinctive styles. Most

Spanish sherries are aged for five or more years, while ours here in New York are rarely matured for more than three years. More significantly, though, the techniques used in their production and aging are different.

"Although only white grapes are used in Spain, both red and white grapes are used here in North America, notably in the East. The wine is heated and simultaneously exposed to oxygen, often by bubbling air through the wine. The heating and air exposure, called processing, oxidize and precipitate any red pigments in the wine. Nevertheless, additional brown pigments are produced during aging. Heating and oxidation also generate the baked, oxidized flavor of these wines. After several weeks of processing, the wine is aged in oak casks and exposed to seasonal temperature variation for several years, occasionally on winery rooftops."

At this point, John again motioned to the attendants to bring another set of glasses. These contained several examples of New York State sherries, and for comparison, two Spanish sherries.

Dave was the first to comment again. "There are distinct differences between the wines presented, especially between the two Spanish sherries. What's the origin of these differences?"

"The two Spanish sherries represent the two ends of the spectrum of Iberian production techniques. The first is a fino sherry–that is, a wine aged in partially-filled barrels, where a thick yeast coating (*flor*) forms on the surface of the wine. During aging, a portion of the wine is periodically removed and blended with older wine, in a precise blending sequence called the solera system. Here in New York we neither promote the production of flor nor employ solera blending. The other sample of Spanish sherry (oloroso) is closer in characteristics to most American sherries. It is not exposed to the action of flor yeasts, but is allowed to oxidize slowly in barrels. The wine is, though, solera blended. The brown coloration and sweetness of the wine come primarily from the addition of boiled-down grape juice–thus, the slightly baked aroma of oloroso cream sherries. Additional sweetness may come from the addition of fortified grape juice, solera aged for various periods. Sweetening of the wine occurs after solera blending is complete."

I was the next to pose a question. "In the production of American sherries, do you use the same grape varieties as employed in Spain?"

"No," responded John. "Here, the choice of grape variety is less important. This results because heat processing tends to destroy the wine's varietal aroma. In addition, the aldehydes and other flavorants formed during oxidation tend to mask varietal fragrances."

"You have commented on the origin of the sweetness found in some Spanish sherries, but you have not indicated where the sweetness of American sherries comes from," said Pat.

"American sherries are almost exclusively sweet," noted John. "This comes from the addition of grape juice to the wine before heat processing and maturation."

"In the samples you gave us, you indicated that they demonstrate the extremes in Spanish sherry styles. I assume, then, that there are several others," Claude said.

"Indeed!" affirmed John. "This also reflects the more diverse uses of sherry in Europe as compared to here. For example, dry to semidry sherries are frequently sampled as an aperitif, taken with hors d'oeuvres, consumed with the meal, or added to food in its preparation as a flavorant. Sweet sherries are typically consumed at the end of the meal with (or in lieu of) dessert. In North America sherry use is often restricted to use as a sweet drink at social gathering, or regrettably, as an inexpensive source of alcohol.

"Although several sherrylike wines are produced in Europe, only one is permitted to be called *sherry*. That is the wine produced, blended, and aged in the region bounded by Jerez de la Frontera and Sanlúcar de Barrameda in Andalusia, Spain.

"Differentiation of sherry styles begins shortly after fermentation. Wines intended for *oloroso* production, such as cream sherries, are fortified to 18 percent alcohol. This prevents the growth of spoilage microbes as well as flor yeasts. These wines are aged in barrels filled to the top. In contrast, wines intended for *fino* production are fortified to only 15 percent alcohol. This creates conditions favorable for flor yeast growth. Filling the barrels 80 percent full provides the surface on which the flor yeasts can grow. The yeast layer formed protects the wine from direct exposure to air.

"All Spanish sherries are blended in a procedure called the solera system, as mentioned previously. It is a fractional blending procedure in which a portion (fraction) of the wine from one set of barrels is added (blended) to the next older set of barrels. Each set of

Production of *Fino, Amontillado,* and *Oloroso* Sherries

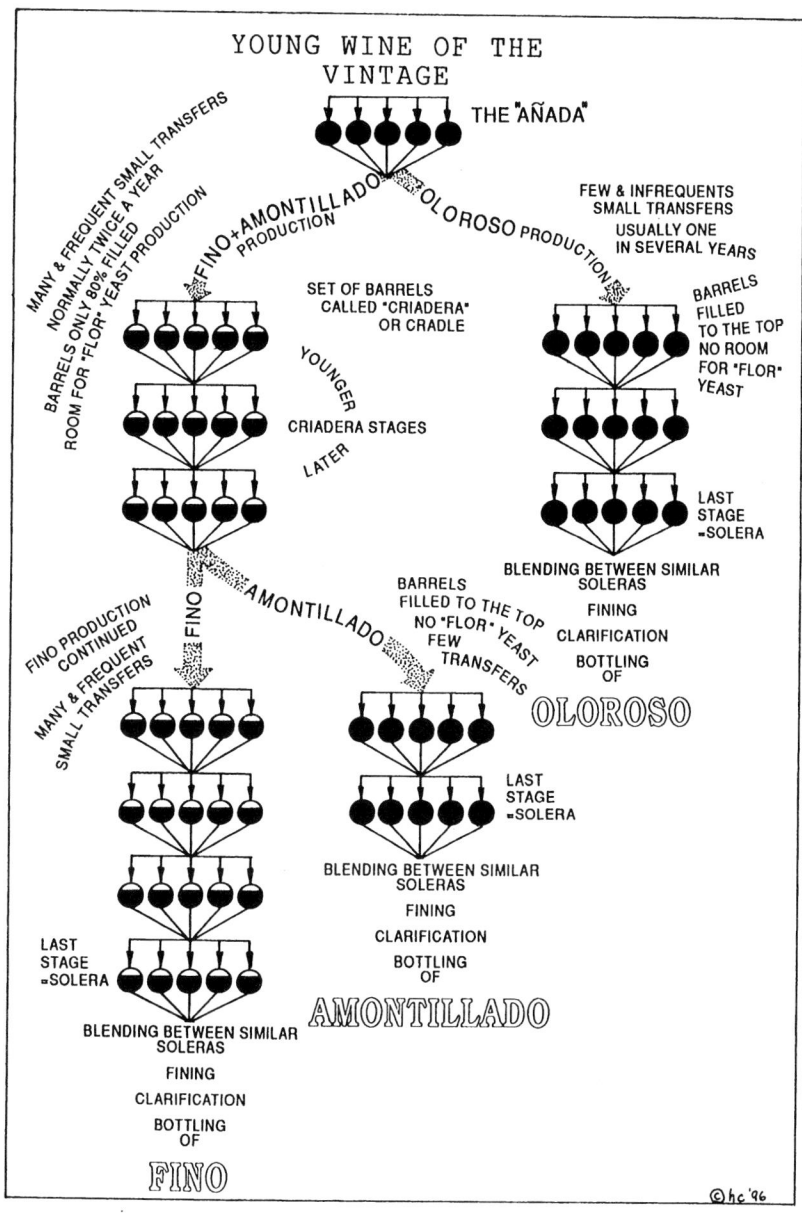

barrels is called a *criadera* (cradle), the oldest of which is called the *solera*. Typically, the sample (~ 20 percent) removed for each transfer is replaced by an equal volume from the next younger *criadera*. The number and frequency of the transfers vary considerably, especially between styles. *Fino* sherries are usually transferred twice a year, whereas *oloroso* sherries may be transferred only once every several years. Typically, *fino* sherries pass through many criaderas, while *oloroso* sherries pass through only a few.

"The frequent transfers used in fino sherry production permits the flor layer to maintain itself. However, if the frequency of transfer is slowed, the flor yeast cap dies and the barrels are filled. This changes the direction of the wine's development, and it progresses toward the amontillado style. Other sherry styles are variations of the procedures used in the production of fino, amontillado, and oloroso sherries. Wine that is removed from the last (*solera*) stage is fined and prepared for bottling."

"In table wines, the best wines are commonly considered to come from single vintages and specific vineyards. This is rarely done in ports and never done with sherries. Could you comment on this?" requested Phil.

"Blending firms (Lodges, Bodegas) have played a major and formative role in the evolution of port and sherry styles," commented John. "In addition, brands are a dominant feature of these styles. Each firm produces its own distinctive versions. Brand consistency is possible only by blending wines across vintages and between vineyard sites. The blending in sherry maturation, as just noted, is highly structured. In port production, the blending schedule is more variable and at the discretion of the cellar master.

"In table wines," John continued, "winemakers have many chances to influence the final taste of the wine they produce, but they are typically limited by the basic traits of the wine produced in any one year. For the sherry and port producer, blending often neutralizes both annual and vineyard differences. This gives the cellar master considerable flexibility to augment or diminish certain features to maintain brand consistency. The use or avoidance of blending to respectively achieve brand consistency or retain vintage and site uniqueness have their separate advantages and disadvantages. Neither is inherently better than the other."

"If sherries do not come from a single vintage, then what is the meaning of the date found on some bottles of sherry?" asked Claude.

"The date refers to when the particular solera system of the wine was started. Typically, only a minuscule fraction of the wine is likely to come from the actual year noted on the bottle. Because of fractional blending, soleras can potentially continue forever.

"Another feature about sherries and ports worth noting is their pronounced flavor," continued John. "This contrasts with the comparative subtlety of most table wines. This reflects the predominant use of fortified wines as appetizers before the meal, with cheeses after the meal, or instead of the dessert. Fortified wines also are consumed in smaller quantities than table wines. Thus, the contents of a bottle may not be consumed completely on opening. The high alcohol content, varying from 17 percent to 23 percent, acts as a preservative, preventing microbial spoilage. In addition, the oxidized nature of most fortified wines prevents rapid changes in flavor after opening. There are a few exceptions to this rule, however. Fino sherries are susceptible to oxidative flavor changes and vintage ports lose their bouquet as rapidly as do table wines on opening."

SPARKLING WINES

Because the spate of questions had dwindled, John turned his attention to sparkling wines.

"It is with particular pleasure that I now turn to one of our specialties at Gold Seal–champagne. Since early in our history, we've been fortunate in attracting several distinguished cellar masters from Champagne. The most famous and influential has been Charles Fournier, who came from Veuve Clicquot-Ponsardin. Charles Fournier brought his own yeasts and set out to enhance the existing reputation of our wines. In addition, he encouraged the cultivation of European (*Vitis vinifera*) grape varieties by Dr. Konstantin Frank. This proud tradition of producing award-winning champagnes and *vinifera* varietal wines has been carried on to the present. Most of our sparkling wines are in the sweeter, fruitier style desired by most of our customers. Nevertheless, we are especially proud of our dry (*brut*) styles, now typical of most French champagnes. Before going any further, however, I want you to try some of the finished product." As if on cue, the attendants began setting six glasses of bubbly in front of each of us.

"While you're sampling the wines, one of which is a Veuve Clicquot from France, a brief history lesson on sparkling wines is in order. I realize that this is not what you came for, but I'm sure you'll find it worthwhile.

"The first sparkling wine to make a real splash came from Champagne, a northern region of France. As has often been the case in the evolution of new wine styles, the end result was neither intended nor expected.

"The wines were also probably sweet. At least, champagne's initial success was with the party crowd in Paris during the early 1800s. This established its continuing association with celebration. However, contemporary connoisseurs disparaged the fashion. How could one, they contended, take seriously this frothy upstart? Champagne's rehabilitation as a product worthy of serious attention began in the latter part of the nineteenth century.

"Initially, wine from Champagne was nonsparkling, thin, acidic, pale colored, and mild flavored. These inherent failings became valuable when producers started to develop drier champagnes. In the process of producing drier wines, prolonged contact with the yeasts embellished the subtle flavors that now epitomize champagne."

TERMINOLOGY ON CHAMPAGNE LABELS

"The slow evolution of champagne, from very sweet to very dry also explains the oddness of some of the terms referring to champagne. *Demi-sec* (half dry) came to differentiate a semi-sweet version, from the very sweet *doux* (soft) style. The subsequent production of a marginally sweet version came to be referred to as *sec* (dry). However, when the first truly dry champagnes were produced, the word dry (*sec*) had already been seconded. The term chosen to designate the dry style was *brut* (rough). Nevertheless, even *brut* champagnes have some sweetness (*dosage*) added prior to bottling. The dosage consists of sweetened aged white wine or brandy. Champagne to which no sweetening is added is called *nature*."

"I have often wondered about the strange terminology found on champagnes," I said. "I had no inkling that it had a historical logic. You also made another interesting comment, concerning the relatively low quality of the wines from Champagne. I'm not clear why this property should be ideal for making sparkling wines."

"The high acidity of the wine gives champagne its lively taste. The wine's mild flavor allows the subtle 'toasty' flavor donated by prolonged contact with the yeast, to express itself. Furthermore, the pale color reduces the tendency of the wine to 'gush' on opening. Tannins extracted along with anthocyanin pigments act as 'nuclei' around which carbon dioxide bubbles can form. In addition, red pigments and tannins disrupt yeast metabolism under the difficult conditions of the second in-bottle fermentation. The sparkle comes from the carbon dioxide released during the second yeast fermentation. Correspondingly, most red sparkling wines possess a lower carbon dioxide content than their white counterparts.

"As the popularity of the sparkling champagne spread, other regions began to experiment with their own versions. One of the early successes occurred in Asti, Italy. With Asti Spumante, the original sweet style of sparkling wines has been retained. The pronounced fruitiness of the 'Muscat' grapes used in Spumante production complements the wine's sweetness. Asti Spumante requires less maturation than champagne because the development of a subtle toasty character is unnecessary. Were the toasty character present, it would be masked by the robust aroma of the 'Muscat' grapes.

"Our own sweet sparkling wines lie somewhere between the fruity spumante and subtle champagne styles. Our sweet sparklers often are produced from the American varietal 'Catawba'. It has little of the so-called foxiness of 'Concord' grapes, but does have a marked floral, almost perfumed character. If we wish to reduce the aroma and diminish the uptake of red pigments, we harvest the grapes before they ripen fully. Rapid but gentle cold pressing produces a pale, mild-flavored juice. This is especially important in our production of subtle-flavored, dry sparkling wines. In the latter style, we often combine wines made from several grape varieties. These may include cultivars such as 'Delaware' and the French-American hybrid 'Aurora.' We also release some pure *vinifera* champagnes, such as our Signature series. These are produced from

'Chardonnay' grapes grown on our vineyards near Valois, on the east side of Seneca Lake.

"Although we occasionally produce a vintage-dated champagne, this is uncommon. Like ports and sherries, champagnes are generally blends of wines from several vineyards and vintages. Wines from several cultivars, as previously mentioned, often help to give the wine the finesse and consistency we want. The quality of the champagne is also significantly dependent on the skill of the cellar master in blending the wines."

PRODUCTION TECHNIQUES: TRADITIONAL, TRANSFER, BULK

"Does this mean that the cellar master makes his or her blend before the wine gets its sparkle?" I questioned.

"Yes," responded John. "After making the blend, called the *cuvée*, a solution of yeasts, nutrients, and sugar, called *tirage*, is added. The *cuvée* is transferred to bottles, capped, and placed in a cool cellar for the second fermentation (about four months). After refermentation, the wines are left in contact with the yeast for a storage period of several months or years. During this time, the yeasts slowly die and release various compounds into the wine. Some of these are important in producing long-lived chains of minute bubbles, and generate the toasty flavor of champagnes.

"After maturation in the bottle, the wine may be released into a large pressurized tank, sweetened, filtered to remove the yeasts, dispensed into sterilized bottles, and corked. This process is called the *transfer* process. More commonly, however, the bottles are placed upside down in large cages. The cages are placed on a mechanical device that periodically shakes the wine. Shaking dislodges the yeast sediment, and inversion of the bottle results in the yeast settling next to the crown cap sealing the bottle. This is the modern version of the traditional hand riddling (*remuage*). The wine is subsequently chilled and the bottle neck plunged into a freezing solution. This causes the formation of an ice plug containing the yeasts next to the cap. The bottles are then inverted, the yeast-containing ice plug shot off following cap removal, *dosage* added, the cork inserted, a wire cage applied, and the bottle labeled. The latter is called the *traditional* or *méthode champenoise* process. In America it is designated by the inscription 'fermented in *this* bottle.' The transfer method is noted by the statement 'fermented in *the* bottle.' The only sensory difference between wines produced using the two techniques is the greater uniformity achieved with the transfer method, due to the blending associated with transfer into the pressured tanks."

"May I interrupt for a moment?" asked Dave.

"Certainly, how can I help?" replied John.

"I hate to appear dense, but what is riddling?"

"Riddling? It's the method most commonly used to remove yeast cells from sparkling wine. Following the second in-bottle fermentation, yeasts settle and stick loosely to the sides of the bottle. In the old days, periodic hand agitation broke up the attached yeast sediment. Positioning the bottle neck downward in an A-frame, resulted in the yeast sediment slowly accumulating on the cap. The procedure often took several weeks or months to accomplish. The modern mechanical procedure can achieve the same result in about a week or two."

"Since the transfer technique seems simpler and produces a more uniform product, why do producers still employ riddling, even if now mechanized?" inquired Pat.

"The answer is in the name." John smiled. "*Tradition*! In the world of wine, it is often more important to give the impression of continuance with time-honored practices than to use improved economy. We have an excellent example with bottle closures. In the second in-bottle fermentation, a crown (pop bottle) cap is used. Not only is it simpler and cheaper, but it is more effective. However, crown caps are used only before riddling, so that consumers won't see them. For the finished bottle, the traditional cork closure is used. The only real advantage I can imagine for the continued use of cork is in sports celebrations. Here, shaping the bottle to induce the cork's projection and spraying everyone in the proximity is 'traditional.'"

"Aren't there other methods of making sparking wine?" questioned Tony.

"Yes," responded John. "The other main procedure, used primarily for inexpensive sparkling wine, is the *bulk* or Charmat process. In this procedure, the *cuvée* (blend) is refermented in a sealed reinforced tank. After refermentation, and a short period of contact with the yeasts, the wine is filtered during transfer into another pressurized tank. The use of a pressurized tank is required, as in the transfer method, to prevent the escape of carbon dioxide. The wine is sweetened and bottled. Although typically used in producing inexpensive sparkling wine, they are not necessarily of poor quality. However, the short time spent on the yeasts does limit the development of a toasty character or the release of yeast by-products required to produce long-lasting chains of minute bubbles.

"There are several additional procedures occasionally used in Europe. However, they are of limited use and appear to have few benefits worth noting. If you really want to know more about these processes, I can suggest some articles you can read."

CHAMPAGNE CORKS

"For the past while I have answered your questions. Now I'd like to reverse the process. I want to ask you a question. We all know the mushroom shape of champagne corks after they come out of the bottle. What do the corks look like before they go in?"

There was a prolonged silence as we passed furtive glances around the table. Phil grinned, indicating that he knew the answer, but wasn't going to spoil John's fun by responding. Finally, Dave admitted ignorance and asked where the cork got its mushroom shape.

John reached down into his vest pocket and said, "You obviously didn't take note of the large containers of oversized corks in the winery." John passed a fat-looking tubular cork around for everyone to see.

"You'll notice that the structure of the cork is different from most wine corks. It consists primarily of agglomerate cork–cork fragments glued together–with two superimposed discs of natural cork bonded to one end. The cork is squeezed and aligned so that, when inserted, the layers of natural cork are in contact with the wine. The upper portion of the cork, not inserted into the neck, is mechanically crushed into the familiar rounded shape. This provides a handle for removing the cork. The flared base of the cork, as it comes out of the bottle, reflects the sloping inner sides of the bottle neck. The widening of the inner diameter down the neck helps hold the cork in place when the wire restraint is removed."

Champagne Corks

©hc'96

"Since you are on the subject, how do you recommend opening a bottle of champagne? I've heard various suggestions over the years," noted Dave.

"My first recommendation is to make sure that the wine is chilled to refrigerator temperature. Once the foil capsule is removed, slowly loosen the wire cage. Once loosened, hold the cage in place with your left thumb. That is, of course, if you're right-handed. While holding the cork and wire cage securely, twist the bottle from its base with your right hand. This breaks the seal between the cork and the neck. Continue twisting the bottle slowly, until you sense the cork starting to come out on its own. At this point, simply regulate the slow emergence of the cork. On reaching the top, there should be a slight *hiss*, as the pressurized gas in the neck escapes. This procedure should assure control of cork removal. Although not generating the celebrated *pop*, you'll avoid wine gushing all over and the danger of a ricocheting cork."

"You have talked about cork closures. What about plastic stoppers?" inquired Phil.

"Plastic stoppers have several advantages over cork. Not only are they less expensive, but they are easier to remove and reinsert, if desired. Cork stoppers are essentially impossible to reinsert. The major disadvantage of plastic is its poorer sealing properties, especially after several years in the bottle. However, as most inexpensive sparkling wines have a high turnover, this failing is not a limiting factor in its use."

"The bottles of sherry and port also need be easily resealed. Why are their corks not plastic?" ventured Claude.

"Tradition, I suspect," said John. "Plastic can be used in inexpensive sparkling wines because the consumers of these wines are unconcerned about convention. In contrast, cork is retained for ports and sherries for market image. However, to permit cork stoppers to be easily reinserted, the bottom rim of the cork is chamfered–that is, trimmed away at a 45 degree angle. Also, the inner diameter of the neck does not increase down the bottle, as usual for bottles of table and sparkling wine. A plastic cap glued to the cork gives the grip needed for easy removal and reinsertion.

"Well, keeners, were you able to detect the French champagne?" challenged John.

It was intriguing to see the diversity of opinion. We all agreed that it was not the first, third, or fourth wines. They seemed both too sweet and fruity for champagne. In addition, although there are some rosé champagnes, no one thought the fourth wine was one of them. Concerning the other three wines, there was no agreement. Phil and I thought that the second wine was the Veuve Clicquot, Tony considered that it was the fifth wine, Claude and Dave voted for the sixth sample, and Pat simply wasn't sure.

Phil and I were overjoyed when informed that we had guessed correctly. John then asked us how we had come to our decisions. Phil's response was that the straw color reminded him of French champagnes, and that the wine seemed to express most clearly a toasty fragrance. My decision was based on it showing the least metallic taste of wines two, five, and six. Generally, I'd found that champagnes showed less of the metallic taste I disliked. Although significant to me, the others in the group couldn't detect the metallic aspect I objected to. Tony said he had chosen the fifth sample because he liked it the most. If one could associate a higher price with the better wine, then the fifth wine should be the most expensive–and therefore, the champagne. It turned out to be the Charles Fournier Select at a third the price of the Veuve Clicquot. Claude chose the sixth wine because he thought it was the most subtle and complex in flavor. Dave thought the sixth sample might be the French champagne because of its color and balanced flavor. Wine number six was Gold Seal's vintage Blanc de Blanc made from its Valois vineyard 'Chardonnay' grapes.

We were all impressed by the quality of the wines. The less expensive, fruitier types were obviously well-made, but simply were not to our taste.

Although this had been a great experience, the plastic and steel seats were getting progressively hard. Also, John had a meeting scheduled with his chief winemaker. John bid us a great weekend and invited us to stay and try some of their table wines. After a round of applause, John disappeared and we were free to sample on.

Except for Pat, who had some experiments to attend to back at the lab, the rest of us availed ourselves of the opportunity to sample a few more wines. I was especially interested in trying their French-American hybrid varietals. Up to that point, I had tried neither a Ravat blanc nor a Rosette. They also had an Aurora and Vidal that I had not sampled before. By the time we were through sampling, Suzanne had returned from scouting the town.

Suzanne sampled those that I thought we might like to have at home, as a prelude to buying. Phil was the only one left when we were ready to leave. We asked him if he wanted to go for a late lunch. Because Gloria was away visiting her sister in Syracuse, he agreed. He knew a small bistro that served light meals just a few blocks away. It specialized in crepes. That was enough to convince us.

Chapter 10

Impromptu Meeting

The timing of our next meeting was quite unexpected. Klaus gave Phil a quick call indicating that a friend had dropped in from Germany. Wolfgang Schmidt had completed his postdoctoral position and was touring the United States on route to Australia. He had accepted a position with the Australian Wine Research Institute in Glen Osmond. Klaus had been telling Wolfgang about our group and he had volunteered to speak to us about his interests and experiences. It seemed like a great opportunity, so we hurriedly set up an impromptu meeting. Because of the short notice, Phil booked one of the meeting rooms in the Statler Inn, associated with the Hotel School.

It was the first really warm night of the year. It smelled like spring. The odor of ammonia and other compounds released from the soil is more a harbinger of spring than hearing the first robin. There was even the occasionally audible drone of a lazy mosquito out on early patrol.

With a name like Wolfgang, I expected to see a tall, blond-haired, descendant of Voltan striding next to Klaus. To my surprise, there was a short, roly-poly, ruddy-complexioned fellow in jeans sauntering toward us. He looked more like a cellar master than a lab scientist. Nevertheless, he was a biochemist turned microbiologist. He had spent the afternoon making a whirlwind tour of several wineries along the way down from Geneva. Although his accent was unmistakably German, his English was impeccable. Why is it that Europeans can master several languages, while we in North America seem to have difficulty being competent in one?

After Klaus introduced us, Phil invited everyone in for a tour of the Hotel School. It was the first occasion for most of us to see the facilities of this famous segment of the university. The Statler Inn is unique in being the equivalent of a teaching hospital allied with a Medical School. It provides direct experience in all aspects of hostelry. Particularly enjoyable was our visit to the hotel's wine cellar. Regrettably, our meeting was not scheduled to be held in such conducive surroundings.

Once we had settled into the cushiony chairs of the boardroom, Klaus gave us a brief biographic sketch of Wolfgang. He was the son of the wine master at Sandgrub, near Eltville in the Rheingau. While in Mainz, he had worked under one of the most famous enological microbiologists in Germany, Dr. Ferdinand Radler. Recently he had been studying at Geisenheim on the origin of wine off-odors. In this regard, Wolfgang had spent several months in Portugal studying the cork industry. In addition, he had been Klaus' roommate while both were attending Johannes Gutenberg Universität.

CORK STRUCTURE AND PROPERTIES

"Thanks!" said Wolfgang. "It's always a pleasure to talk about wine. As Klaus stated, I have been most recently involved with off-odors, particularly those derived from cork. While in Portugal, I had an opportunity to investigate the harvesting of cork and its processing into stoppers.

"Cork is the outer portion of the bark on trees. The bark of the cork oak differs from most trees in its thickness. In addition, its bark can be of uniform structure. I say 'can be' because the cellular homogeneity develops only after the bark has been stripped off the tree several times. The first two strippings, called *virgin* and *second* cork respectively, are too cavernous and variable in structure to produce quality stoppers. Somehow, stripping stimulates the tree to regenerate bark of a more consistent cellular structure. The third and subsequent strippings, called *reproduction* cork, possess the physical and chemical properties required for cork closures. It takes about nine years for the tree to produce a sufficiently thick layer of new bark for the production of cork stoppers. The trees are stripped on a nine- to ten-year cycle for the next several hundred years.

"The next time you open a bottle of wine, look carefully at the cork. You'll notice faint lines running down and across the ends of

Cork Stoppers

©hc'96

the cork, if it isn't agglomerate cork (composed of cork chips glued and pressed together). When counted, you'll usually find about seven to nine lines. These correspond to the annual growth rings of the cork, equivalent to the growth rings found in wood. The growth ring pattern in cork also reveals that stoppers are punched out down the grain of the bark."

"Is there any particular reason why corks are stamped out in this manner?" I asked.

"Yes, several reasons," responded Wolfgang. "From the point of view of the manufacturer, it permits stoppers of different lengths to be obtained from the same slab of cork. However, there is a less obvious, but even more significant reason. This concerns the sealing property of the stopper.

"On the trunk, cork insulates the tree against damage caused by fires that historically ravaged western Mediterranean forests. Although an effective insulator, this property severely retards the diffusion of air into the trunk. To permit gas exchange between living cells in the trunk and the atmosphere, pores–called lenticels–span the cork."

Wolfgang turned to Phil and asked if he could get some corks for people to see what he was talking about. In short order, Phil was back from the kitchens with his cupped hands full of corks.

Wolfgang continued. "If corks were cut out radially across the bark, the lenticels would run the length of the stopper, making them leaky. As it is, the lenticels run across the stopper, their ends abutting against the bottle neck.

"Note also the variable quality of the corks. Some have large lenticels (milk-chocolate colored in cross-section), as well as cavities (gaps or cracks). The latter may be filled with cork dust and glue. Some of the corks also have been stained or heavily bleached. Natural cork is beige-brown in color. See if you can find any corks with more than nine growth rings."

Cork Bark with Stoppers Punched Out

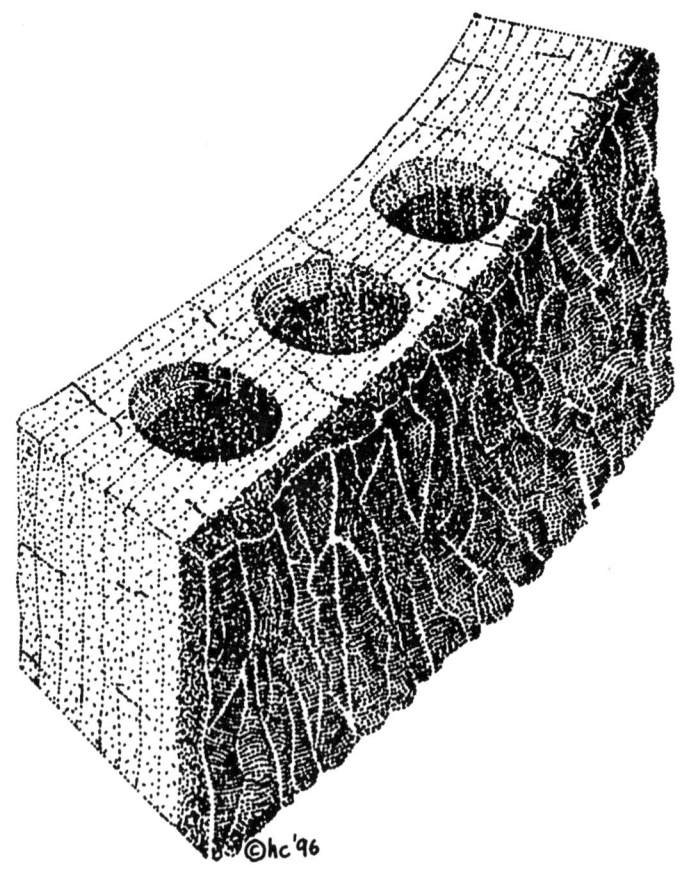

Pat was the first to find one. "What does the greater number of growth rings signify?" he asked.

"Corks with many, closely-spaced growth rings probably came from mountain-grown cork," noted Wolfgang. "Because trees in mountainous regions grow more slowly, the growth rings of the bark are narrower. Thus, it may take twelve or more years to produce sufficient cork for stripping. Of greater importance, though, is the greater elasticity, and therefore better sealing properties, of the cork."

"What is the source of the shiny, slippery surface on some of the corks?" I asked.

"Oh!" exclaimed Wolfgang, half-laughing. "That's due to the addition of paraffin or silicone to ease the insertion of the stopper. It tends to avoid the creases (and leaks) associated with corks slightly off center when inserted into the bottle."

CORK-BORNE FAULTS

"Since we are into the appearance of cork, what's the cause of the black material one often finds on corks in old bottles of wine?" inquired Dave.

"In most cases it develops from mold growth forming on wine that may have seeped out through gaps between the cork and neck of the bottle. These gaps usually seal within a day or two, as the cork completes its enlargement after being squeezed for insertion. To prevent such wine seepage, bottles are usually left upright for about twenty-four hours after filling. Alternately, wine may ooze out following rapid increases in wine temperature (and volume), seep out through internal cork faults, or escape due to the loss of cork elasticity with age. Under metallic neck capsules, humid conditions often favor the growth of fungal contaminants. Subsequently, the molds die, oxidize, and turn brown to black, along with wine residues."

"Is this dangerous?" continued Dave.

"By itself, no. However, the production of fungal acids, and the acidity of wine residues, can dissolve lead from tin/lead capsules.[1] The lead salts formed can produce a crust on the top of the cork and

---

[1]This does not occur with aluminum or plastic neck capsules.

lip of the neck. These salts, if present, can easily be removed by wiping the neck of the bottle with a damp cloth."

"Will lead salts not diffuse through the cork into the wine after many years?" asked Phil.

"The evidence to date suggests no, even though cork is permeable to water. The loss of water from the cork explains why wine bottles need to be stored on their sides. If the cork dries, it shrinks. Even more important, dry cork loses its elasticity. This reduces the cork's ability to press tightly against the neck. Stoppers are always slightly larger in diameter than the neck of the bottle into which they are inserted. This is essential to assure that the cork presses snugly against the sides of the neck."

"Getting back to lead, is lead contamination really a problem?" pursued Claude.

"No. The concentration of lead in wine is very low, often lower than in other foods. Lead contamination could be a problem, though, if one were to store wine for extended periods in lead crystal decanters."

"Thus far, I haven't talked about off-odors in wine," continued Wolfgang. "Because cork cell walls are waxy, stoppers can absorb odors from the surrounding environment. Subsequently, these compounds may contaminate the wine."

"Is this the origin of most corky off-odors?" questioned Pat.

"In fact, no, even though I happen to have mentioned it first. Most so-called corky off-odors probably originate from excessive exposure to bleach during processing. Manufactures use chlorine bleach both to surface-sterilize and to decolorize stoppers. Low quality corks are more susceptible to chlorine contamination because of their large lenticels and crevices. These faults permit the chlorine to seep further into the cork, and may not be adequately removed during rinsing. Moldy odors form through a complex series of chemical reactions. Similar compounds also may form due to the use of chlorine-containing pesticides on cork trees. In addition, wood structures treated with chlorine-containing pesticides may be an important source of moldy, corky odors. In damp, poorly ventilated cellars, these compounds can be adsorbed by barrels and subsequently contaminate the wine stored therein."

"It sounds as if cork has been given a bum rap," asserted Pat.

"I do not know what a 'bum rap' is, but if you mean that most so-called corky odors are not directly the fault of the cork, the answer is yes. One of the benefits of cork is its basic freedom from odor. Cork stoppers were used as early as the fifth century B.C. by the ancient Etruscans to seal wine amphoras."

"Excuse me," Dave interjected. "What are amphoras?"

"Amphoras are relatively large (five to ten gallon) ceramic vessels that look like immense truncated carrots or plump onions. They often possess two handles near the thick neck, and have a roughly pointed bottom. Amphoras were extensively used in ancient times for the storage and transport of various dry and liquid goods. Early versions were often coated on the inside with pine pitch to make them water-proof for liquid storage. Subsequently, potters were able to glaze the amphoras sufficiently to eliminate this necessity."

"Would not the use of pitch have spoiled the taste of the wine?" said Dave.

"To our taste, and that of some Roman authors such as Pliny, definitely. If you want a hint of what wine from pitched amphoras

Wine Amphora

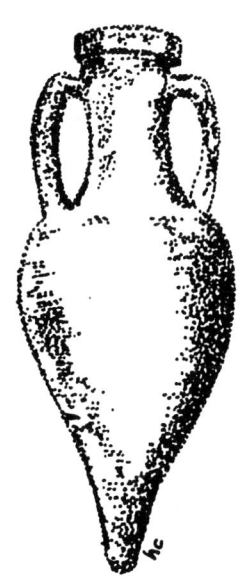

might have tasted like, try a bottle of Retsina. Greece is the only country that retains a liking for wines with a piny essence."

"So that I don't get it wrong, are you saying that natural cork is essentially flavorless, but can become contaminated with smelly compounds or from excessive bleaching?" inquired Dave.

"That, at least, was what I was trying to say," responded Wolfgang.

"If cork and oak are even indirectly the source of most corky or moldy odors, why are they still used?" inquired Pat.

"That is a complex question with no simple answer," responded Wolfgang. "Part of the answer is linked with their traditional use. Cork probably would have been replaced by metal caps long ago, were it not for consumer resistance. Conversion to metal caps, such as used on inexpensive wine, would not influence quality at all. In fact, aluminum roll-on closures do a better job at excluding air and odors, are less expensive than cork, easily reseal the bottle, and allow wine bottles to be stored upright."

"So it's not poor quality that has limited the use of metal caps on expensive bottles of wine?" proposed Claude.

"Correct! There actually is no evidence that cork stoppers are necessary for proper wine aging. Old views, based on little or no factual evidence, are repeated so often that they acquire the aura of truth. Take, for example, the oft-repeated statement that wine 'tears' climb up the glass. Pour some wine gently into a glass and wait for tears to climb. You will be there, as you say, 'until hell freezes over.'"

"Can one tell by smelling the cork whether the wine is likely to be contaminated?" asked Claude.

"Often, but not consistently," said Wolfgang. "I remember once smelling a cork with visible fungus growing on its sides. It had a corky odor, but the wine itself was faultless. Nevertheless, the common situation is where the cork itself smells alright, but the wine has an off-odor."

TYPES OF OAK:
STRUCTURE AND PROPERTIES

"Well, on to oak," continued Wolfgang. "Although the bark from the cork oak is used because of its neutral flavor, the wood of other oak trees is used to give wine a mild woody flavor. Its association with some regional wines is so strong that oak flavors are considered an essential ingredient in the wine's fragrance.

"In Europe, two oak species are used in barrel construction, while in America several different but closely related oaks are used. Nevertheless, all are white oaks; that is, their wood becomes almost impermeable to liquids and gases as it matures. Both American and European white oak produce tight (nonleaky) barrels and other large cooperage. Thus, a preference for cooperage made from American or European oak depends on other properties. For example, American oak releases fewer tannins than European oak, though both liberate about the same amount of aromatic compounds. However, other factors can be more important in affecting the wine's flavor than the oak species used. These include the local forest climate, the tree's growth rate, the site in the log from which the wood came, the method and duration of seasoning, and the degree of toasting and method of wood softening used in barrel manufacture. Each of these factors subtly modifies the chemistry of the wood, and the flavors released into the wine.

"Of these factors, the most well-known is the preference of certain regions, such as Bordeaux, for barrels made from the oak of particular French forests. This is an example of selection for the influence of local climatic conditions. What is less well appreciated is that the preferred forest is usually the closest to the wine region concerned. Proximity and habituation are often the mother of traditional preferences."

"Your latter observation has wide applicability, and pertains to the supposed combination of local wines with regional cuisines," added Phil.

TOASTING

"Could you explain your use of the term *toasting*?" questioned Pat. "Several weeks ago we heard about the desired toasty flavor of champagnes, and now we have toasty oak. What's the connection, if any?"

"Toasting is a term used to describe the effect of fire on wood chemistry. In traditional barrel making, partially constructed barrels are placed over an open fire to relax the wood's fibers. This reduces the likelihood of cracking when the staves are bent into their curved shape. The heat also helps fix the wood in its bowed shape. Depending on the specifications of the winemaker, the period of heat fixing (firing) can be short or prolonged. A few minutes of heat fixing produces only a mild caramelization and browning of the inner wood surfaces. Prolonged firing produces the charred inner surface, commonly used for maturing bourbon and several other distilled beverages. Moderate firing, of course, produces intermediate changes in wood chemistry."

"What are the relative advantages of the various levels of toasting?" inquired Tony.

"With untoasted staves, the wine gains a natural oaky flavor and extracts the maximum wood tannin content. This is where the wood is steamed, rather than exposed to fire, to soften the staves for bending during barrel construction. Light toasting adds a subtle roasted note to the natural wood flavors. Medium toasting reduces the natural oak flavors and replaces them with a vanilla/roasted flavor. Strong toasting further diminishes the natural oak flavors and extraction of tannins, reduces the vanilla and roasted character, but increases the presence of smoky, spicy flavors. These influences are most marked during the first filling of the barrel, and progressively change in strength and character during repeated use."

"This story of oak sounds pretty complicated. How does a winemaker decide what to do?" inquired Claude.

"Experience!" responded Wolfgang. "The winemaker must know the properties of his wine, his consumers' preferences, and have a good palate. To enhance the subtlety of each batch, the wine may be aged in barrels with different degrees of toasting. The cellar master periodically samples each barrel and determines how long the wine should be matured and in what proportion they should be blended. The decisions are often empirical."

"If the flavors provided by barrels decline and change with repeated use, how often can a barrel be used?" Pat asked.

"It really depends on the intentions of the winemaker, unless dictated by appellation control laws. Because of the expense, only a portion of the wine is usually aged in new barrels. This wine is subsequently blended with similar wine aged in used barrels. This is not necessarily a disadvantage. Exclusive use of new barrels can provide the wine with an excessive dose of tannins and mask grape aromas."

"Does the extraction of compounds from the wood weaken its structure and limit the number of times a barrel can be used?" asked Claude.

"No, the structural integrity of the wood is not compromised by exposure to wine. What can weaken the barrel is shaving its inner surface. This technique exposes inner tissues of the wood to wine. Wine only extracts material from the innermost few millimeters of wood. Because the wood exposed by shaving was little affected by the original firing, the barrel may be refired to give the newly exposed surfaces the desired degree of toasting."

"Would not oak chips or shavings, toasted as desired, give the same effect, but be considerably cheaper than using barrels?" asked Pat.

Wolfgang smiled. "Some people are doing just that. They add oak chips during maturation in stainless steel tanks. Apparently the procedure works well. An alternative is to add strips of oak, toasted to the desired degree, to tanks of wine. However, one feature is usually missing with either technique. During normal barrel aging, the wine is slightly exposed to oxygen. This both helps to stabilize the color of red wines and produce a slightly oxidized character. Some winemakers like the latter feature. The exposure to oxygen comes during racking (when the wine is removed from the sediment), sampling, or topping of the barrels. Topping was once considered necessary to replace the wine lost by evaporation through the wood. Air was thought to fill the ullage space and favor wine spoilage. However, if sealed properly, air does not seep into the barrel, and a partial vacuum actually forms over the wine."

IN-BARREL FERMENTATION

"On the label of some wines, one now sees the phrase *in-barrel fermentation*. Is this just a marketing ploy or does it really make a difference?" probed Klaus.

"*In-barrel fermentation* refers to the production of wine in small (∼50 gal) wooden barrels, rather than the more usual large fermentation tanks (>1000 gal). This difference does subtly influence the wine's character. Whether one prefers the difference is, of course, a matter of choice. Wine fermented in-barrel may be spicier, oakier, smoother, and possess a less buttery flavor than its tank-fermented version. In-barrel fermentation is more commonly used with white wines than with red wines. Tannins extracted from the wood tend to adhere to yeast cells, and thus precipitate out of the wine in the lees. This can make the wine smoother tasting than the same wine fermented in a tank, and subsequently matured in oak barrels. However, because oak tannins also promote the early loss of anthocyanin pigments, red wines are rarely fermented in new barrels. Desirable oak flavors are not equivalently affected, because they are extracted more slowly. Thus, oak flavors accumulate only if the wine is also barrel-matured."

AGING SUR LIES

"Is in-barrel fermentation related to what is termed *sur lies*, found on some wine labels?" inquired Claude.

"Aging *sur lies* is not directly related, but may follow, in-barrel fermentation," responded Wolfgang. "*Sur lies* refers to the maturation of wine in prolonged contact with the lees (sediment) that settle out during aging. During in-barrel maturation, the wine is periodically stirred. As a consequence, the wine is exposed to air and absorbs oxygen. This helps limit the production and accumulation of sulfur off-odors. However, oxygen uptake can favor the growth of acetic acid bacteria and the potential development of a vinegary flavor. Although some producers have success with the *sur lies* process, and believe that it improves their wines, most winemakers don't consider it worth the risk."

"Because Wolfgang has had a long day, maybe we should have a short break and sample some wine," Phil interjected. "When picking up the corks, I made a quick check of our cellar list. On it, I found a couple of bottles of Erbacher Sandgrub Riesling Auslese. I thought it would be fitting for us to sample the wine in the presence of someone so closely linked with the property. Were you involved in its production?"

"As far as being a grape picker, probably," responded Wolfgang. "As far back as I can remember I have been picking grapes."

"I admire the logical organization of information on German wine labels," I noted. "However, why are not the names of individual vineyards easily distinguishable from group vineyard names. Because individual vineyards may produce more distinctive wines, it would be useful if one could easily tell them apart without having to resort to the *German Wine Atlas*."

"It is unfortunate that the 1971 Wine Law did not make this difference clear. Although not compensating for this lack of precision, our labels do always mention the region from which the wine comes. This is more than one can say for some other European wines. I realize the importance of knowing a wine's geographic origin to consumers outside Europe. In Europe, most people drink wine produced only from their own area. Thus, there is little need to indicate the wine's geographic origin on the label."

CLASSIFICATION OF GERMAN WINES

"While we're awaiting the arrival of the wine," Dave said, "would you mind briefly covering the major terms on a German wine label?"

"It would be a pleasure to talk about my country's wines," commented Wolfgang, sitting upright in his chair. "With the exception of our most inexpensive wines, the label commonly lists in sequence the town, vineyard site, grape variety, and quality designation. Also noted are the producer, the vintage, and the wine's AP number.

"The AP number refers to the test the wine must pass to use one of the quality designations I will mention shortly. Several bottles of the wine are held in storage in case complaints arise about the wine. If so, the samples can be taken out for analysis. However, maybe I should go back and discuss the official name of the wine.

"The *name* of the wine is its geographic origin. First comes the name of the town with which the vineyard is associated. Origin is designated by the suffix *-er*, meaning *from*. You should recognize its equivalent use in expressions such as "New Yorker." The second word is typically the name of the vineyard, or a group of adjacent similar vineyard sites.[2] If the wine is made from a single grape variety, its name appears next. This is followed by the quality ranking of the wine. These include the basic *qualitätswein*, and premium *prädikat* designations. The prädikat category is subdivided into six grades: *kabinett, spätlese, auslese, beerenauslese, trockenbeerenauslese,* and *eiswein.*

"The prädikat grades designate the maturity and characteristics of the grapes used in making the wine. Kabinett indicates fully mature grapes. Spätlese is used for mature grapes, picked a week or more after the designated kabinett harvest date. Auslese wines are made from specially selected clusters of mature, late-harvested grapes. Beerenauslese (BA) wines, as the name suggests, are produced from individual berries (*beeren*), specially selected (*auslese*) for their superior ripeness. Trockenbeerenauslese (TBA) wines are derived from dried (*trocken*) berries (*beeren*), specially selected (*auslese*) for superior quality. The fruit used in the production of auslese, BA, and TBA wines are typically affected by noble rot (*edelfäule*). Nevertheless, under dry, sunny, autumn conditions, grapes may shrivel sufficiently to permit the production of the top categories without the influence of *Botrytis cinerea.* The final category, *eiswein*, comes from frozen grapes, as the name denotes. Because the grapes are left in the vineyard until a hard frost, they are typically protected by netting from the ravenous appetites of birds. Freezing concentrates the soluble components of the grapes, which escape from the ice and grape pulp during pressing. In this regard, I was interested to learn that conditions here and in Ontario often permit the production of your own equivalent of an *eiswine.*

"BA, TBA, and eisweins are always sweet. The sugar concentration of the fruit, cool temperatures, and other factors usually limit the yeast's ability to ferment the sugars completely. Auslese wines are typically sweet, but occasionally may ferment dry. The sweetness that typifies kabinett and spätlese wines usually results from the addition of a portion of the harvest (unfermented juice) to the finished wine. If the wines are left dry or semidry, they are normally identified by the terms *trocken* and *halbtrocken*, respectively."

---

[2]For a few famous vineyard sites, the name of the associated town is not found, for example Scharzhofberger (in the Mosel) and Schloss Vollrads (in the Rheingau).

While Wolfgang had been talking, Phil had located the wine and poured it into glasses. As we sampled the wine, Tony asked if vineyards had to be harvested repeatedly for each category.

"Certainly not where I come from, nor is this typical of other vineyards. Not only is multiple harvesting unnecessary, but it would be exorbitantly expensive. If desired, portions of the vineyard are harvested at different times. One portion may be picked early to produce *qualitätswein*, as a hedge against subsequent bad harvest weather. However, if the weather seems promising, most of the grapes are picked after the officially set harvest date to produce *kabinett* wines. Assuming the weather continues to hold, a further section of the vineyard may be left, and picked later to produce higher grade *prädikat* wines. By varying the degree of selection of the later harvests, one can make wines covering the range from spätlese to TBA."

"This means that a vineyard could produce all categories in a single year?" I questioned.

"Theoretically yes, but this seldom happens due to the complexity of fermenting each fraction separately. Nevertheless, in good vintages, several categories are commonly produced. However, in poor years, the whole crop may attain only the basic *qualitätswein* grade. This illustrates a unique feature of German wine labelling. German wines must annually earn the right to use the quality designation they bear on the label. This is in marked contrast to assigning quality based on vineyard origin, regardless of the conditions during the vintage.

"Now that you have had a chance to try my father's wine, what do you think of it?" Wolfgang chuckled. "I should not have done that, but it's fun to put people on-the-spot."

"As our group's most confirmed devotee of botrytized wines, I must congratulate your father on the superb wine" I stated. "What I relish in ausleses is the combination of varietal aroma with a rich honey/apricot *Botrytis* fragrance. Combined with an immaculate balance between sweetness, acidity, and flavor, such wines are truly seductive."

"And it tastes good too!" gibed Pat.

"OK, so it has sophisticated assertiveness, presumptuous breeding, crisp authority, complex balance, elegant power, and respected finesse. What's it taste like?"

After a round of mock applause for Pat's perceptive remark, Wolfgang thanked us, *both*, for our kind comments. He also volunteered that he was especially pleased with the wine himself.

BOTRYTIZED WINES

"I don't know to whom I should direct this question?" commented Claude. "Both you and Ron may be able to answer it. What gives botrytized wines their distinctive flavor?"

"Because the question is biochemical, Wolfgang is undoubtedly better qualified than myself," I replied.

"I'm not sure anyone is able to answer the question," admitted Wolfgang. "What we presently know is that the fungus involved, *Botrytis cinerea*, releases several enzymes. These destroy compounds that give grapes their distinctive aroma. Depending on the extent of this action, wines produced from infected grapes may possess little or no varietal aroma. However, a distinctive botrytis fragrance replaces those aroma compounds destroyed in the grapes. The most significant of these may be sotolon (4,5-dimethyl-tetrahydro-2,3-furandione). I don't know if that really answers your question?"

"Another aspect of the story," I added, "is the concentration of compounds during infection. Foggy nights followed by dry sunny days retard growth of the fungus, but promote berry water loss. The net effect is a concentration of sugars and aromatic compounds. In addition, fungal consumption of acids results in only a moderate increase in relative acidity. Finally, *Botrytis* produces glycerol that contributes to the smooth sensation of the wine."

"From the discussion thus far, people might think that *Botrytis* is one of the winemaker's best friends. This, however, is not so!" remarked Klaus.

"You're right!" I replied. "*Botrytis cinerea* is more a malicious Mr. Hyde than it is a beneficent Dr. Jekyll. Without daily drying, infected grapes rapidly turn into an ignoble mush. The climatic conditions under which noble rot develops are specific and seldom found. Thus, those few regions where conditions are conducive to noble rot development are justly famous."

"If we take, for example, my own region of the Rheingau," noted Wolfgang. "Cool air flows down the slopes and out over the warm waters of the Rhine. This generates nightly foggy conditions along the riverbank and partly up the slopes. In the morning, the warm autumn sun burns off the humidity, the grape surfaces dry, and the fruit continues to ripen."

"Do the grapes rot uniformly under these conditions?" inquired Dave.

"Rarely," responded Wolfgang. "Often healthy, as well as slightly, moderately, and heavily infected berries occur in the same cluster. Immature, greenish berries are relatively resistant to infection, but become increasingly susceptible as they ripen and turn yellowish golden. This explains why whole clusters may be used to make an auslese, but only individually selected, shriveled berries can produce a TBA."

"In class," noted Phil, "I mention that eisweins typically are not botrytized. I am correct in that, am I not?"

"You are correct," confirmed Wolfgang. "Because the grapes stay on the vine so long, often into December or possibly January, botrytized grapes would probably succumb to other fungi, bacteria, or insects, making them unusable."

"Because of the obvious risks involved in delaying harvest to make botrytized wines or eisweins, I assume that they are expensive?" said Pat.

"Regrettably for consumers, yes," responded Wolfgang. "Not only is there the danger that no wine may be produced, but also the volume of wine made is drastically reduced. Even fermentation of the concentrated juice is tricky."

RECIOTO WINES

"All this is nice, but it seems that thus far you have been talking only about white wines. As a lover of red wine, are there any botrytized red wines?" said Tony.

I was just about to reply in the negative, when Wolfgang answered in the affirmative. I was so stunned by Wolfgang's comment that I sat silent, awaiting his response.

"Until a short while ago, I would have said no," continued Wolfgang. "*Botrytis* is known to produce a potent enzyme oxidizing grape pigments. Thus, red grapes infected by *Botrytis cinerea* produce a brown wine. . . . Typically, that is!

"While attending a conference in Asti, Italy last year, I met Dr. Luciano Usseglio-Tonasset. During our discussion, he mentioned work he had done on Recioto Valpolicella wines. I was astounded to hear that the grapes were botrytized."

"But what about the effect of the enzyme laccase on the red pigments?" I argued.

"I do not know," responded Wolfgang. "The grapes look healthy when harvested. Infection progresses very slowly during the months the grapes are stored on trays under dry, cool, shaded conditions. Even after several months' storage, the grapes rarely produce the feltlike fungal covering usually seen on field-infected grapes. Somehow, enzyme production is severely limited, or the enzyme is largely inactive under the conditions of storage. In either case, the wine produced is red, but show many of the diagnostic chemical attributes of noble-rotted grapes."

"A-ha!" said Tony. "I have always felt that Italian winemakers had great techniques used only in Italy. Although some young winemakers in Chianti disparage the *governo* process, it was instrumental in the initial popularity of Chianti worldwide. Also, who else makes *vino santo?* Now I realize that we developed the unique *recioto* process."

"For those less familiar with Italian wine-making procedures," I inquired, "what is the *governo* process? Also, I don't know what a *vino santo* is."

"The *governo* process involves keeping some of the grapes aside from the main crush," responded Tony. "The grapes are subsequently crushed and added to the newly fermented wine. This induces a second fermentation that, by unknown means, speeds the wine's maturation to drinkability.

"Now on to *vin santo,*" continued Tony. "*Vin santo* is a wine produced from grapes dried for two to four months; the juice derived is placed in small barrels for a slow fermentation in attics. Here, the wines develop under the seasonal extremes of heat and cold for two to six years. It produces a unique dessert wine taken only on special festive occasions."

"I have been reading about Italian wines recently," commented Phil. "However, I've never seen anything even hinting that recioto wines, such as amarones, are botrytized!"

"Don't be surprised," consoled Wolfgang. "Hardly anybody else knows either. There is only one scientific paper on it, and that's in Italian."

"In that case, I think I must raid the wine cellar once more," commented Phil. "We should toast, by tasting a Recioto della Valpolicella Amarone, our discovery of its botrytized nature."

In a flash, Phil was back clutching a bottle of the precious red liquid.

After pouring the wine, Phil was the first to speak. "I had always considered that the distinctive fragrance of amarone wines came from the grapes–'Rondinella,' 'Corvina,' and 'Molinara.' On thinking back, I remember having had wines from Lombardy that tasted and smelled similar to Amarones. However, they were made from 'Nebbiolo' grapes."

Looking at Wolfgang, Phil asked, "Did the article mention any other regions where the same technique is used?"

"I cannot remember for sure." Wolfgang pondered. "However, it may have mentioned Lombardy. If so, it would correlate with your recollection of those Lombardian wines."

"In looking at this wine," I commented, "it is easy to understand why people would not suspect it to be botrytized. Although somewhat brickish for its age, the wine is still decidedly red. Also, the fragrance does not resemble any white botrytized wine I know of. It seems to have a sharp, spicy, phenolic fragrance that in some ways reminds me of the smell of tulips, or maybe daffodils."

"I've heard of wines smelling of almost everything," remarked Dave jovially, "but this is the first time I've ever heard of tulips and daffodils. I don't think I've ever even smelled a tulip."

"That shows you're not a botanist!" I quipped.

Since it was now getting late, Phil called an end to our "formal" meeting with an expression of gratitude to Wolfgang for taking so much of his short visit to share his knowledge of wine with us. While Klaus took Wolfgang to his hotel, the rest of us lingered, as usual, sampling wines and discussing the information we had gained.

Sabbaticals are strange and wondrous events which often precipitate dramatic changes in one's life. Although I had originally gone to Cornell to study the genetics of *Botrytis*, these unplanned, unanticipated meetings were turning out to be more important and stimulating than my initial purpose in going there.

Chapter 11

Visit to the Geneva Research Vineyard

Having completed my sclerotial count in the greenhouse early, I decided to wander toward the Campus Store to do book browsing. I was elated at my success in getting the sclerotial structures of *Botrytis* to germinate. It was becoming a joke around the department that I would soon be entering the gourmet food business with my fungal cups. However, on the way, I made a detour to drop in on Phil. I wanted to check on the time we were to assemble at the Geneva Research Station.

Because Phil was occupied with a student, I took the opportunity to look at the display cases on the wall. One that really took my fancy was a set of incredibly lifelike ceramic mushrooms. To my surprise, the artist (Ernst Lorenzen) lived in Nova Scotia, not far from St. Francis Xavier University, where Suzanne had gone to study English. Before I could read much more, the student left Phil's office. As I entered, Phil was preparing to leave for home. Because of the hour, he asked if I wanted a lift. Since I needed nothing from my office or the Campus Store, I accepted his offer.

While heading for his car, Phil asked how my weekend had been. I rolled my eyes heavenward, indicating it had been "different." Phil laughed and insisted that I relate the adventure.

It had all started simply enough. Doesn't it always? During the previous week, Bruce and Barb (my brother and sister-in-law) had stopped in on their way to the Carolinas. They had brought mother along for a visit. Because the weather was so inviting, it seemed a pity not to get outside. I was keen to return to a wooded area where I had been the previous fall with Dr. Korf's mycology class. I also thought that I might combine our outing with a fungal foray. I should have known better. My collecting trips are often restricted to a few square feet, the time being spent crawling on all fours, or lying prostrate on the ground. With forceps in hand, one painstakingly turns over an endless array of dead leaves and twigs, rummaging for miniature mycological marvels–the brownish cups that arise from sclerotia. Although Suzanne says sclerotia remind her of mouse droppings, I see no such zoological resemblance. Normally I conduct my foraging well off the beaten path. Otherwise, the looks you get from passersby can be priceless! They don't know whether to ask about your state of health, ignore your sprawling form, or beat a hasty retreat. It must look like a simpleton playing pick-up-sticks with the forest litter.

This time, though, I walked like a normal person. Only occasionally did I drop to the ground, appearing to pray for some anaemic marsh marigold, chlorotic trillium, or wilted dogtooth violet. My invocations were of no avail, as I repeatedly came up empty-handed. Thus, it was *sans* fungal treasures that we returned to the wagon. Instinctively, I began to search for the keys in my right coat pocket. No keys! They must be in my left pocket. Not there–rots! Where did I place them this time? A search of my pants pockets retrieved only a used Kleenex tissue, some crumpled leaves (with interesting spots), a broken elastic band, several apple seeds (that I must have stored on my way to a garbage pail), and assorted bits of lint. With my adrenalin level starting to rise, it was time to do the frantic self-frisk to locate those keys. Nothing! Had I given them to Suzanne or to Mother? No! It is at such moments that my outwardly calm nature tends to undergo a transformation. No keys to the car, the apartment, the university, seven miles from the nearest habitation, I am no longer a happy camper!

Our original excursion had been interrupted only by my occasional mycological quests. Now we probed, jabbed, fished, and scuffed our way through the accumulation of dried leaves and twigs scattered along the path. While I was full of self-indignation, Suzanne decided we needed divine intervention. I don't know what she promised Saint Jude, but it must have been adequate. I was just about to give up, when, turning my head, I spotted the glint off something shiny in the distance. I bolted, oblivious of the hummock pools separating me and the marsh marigolds, under which I had been searching earlier. At my soggy feet, smiling back at me like truant toddlers, were those errant keys. The prodigal son couldn't have been greeted with more joy. In the evening we had a key celebration. I pulled the cork out of one of the Barolos I'd been saving for a rainy day. Rainy or not, there seemed no more auspicious occasion to make joyous the heart of man . . . and women too!

By this point in the story Phil was approaching a good spot for me to disembark. Phil thanked me for his laugh-of-the-day. I responded with a silent prayer that I may never have another occasion of similar ilk to recount. Just before I closed the car door, I remembered that I hadn't asked the question for which I'd gone to his office in the first place. Our rendezvous at the research station was scheduled for 10 A.M.

This was the second visit for me to the Geneva Research Station, but the first for most of our members. I had been there the previous fall to give a talk on some of my previous research on *Botrytis convoluta*, a destructive iris pathogen.

Klaus was waiting for us at the station with another researcher, Rod Vine. We all thought he had a most appropriate name for a viticulturalist. Klaus had asked Rod to accompany us, as he was directly involved in the ongoing research at the station. After a brief introduction, we all piled into one of the station's vans for the short trip to the research vineyards.

When we arrived, Rod gave us an official welcome to the Research Station. As we headed off toward the vines, Rod explained that both he and Klaus would relate different aspects of the studies being conducted at the station. If things were not clear, we should not hesitate to ask questions. They were accustomed to grilling from grape growers who came to see their work. Also, we should feel free to ask questions about any other aspect of grape culture.

ROOTSTOCKS

"The first plot you see to your right," Rod began, "is a trial on different rootstocks. The fruiting part of most grapevines, called the scion, is grafted onto the root system of another vine, termed the rootstock. If you grow fruit trees at home, you'll probably know the shoot is grafted onto a dwarfing rootstock. Similar rootstocks are used in commercial orchards to limit tree size. Although most grapevines are grafted, dwarfing is not the purpose here."

Kneeling next to the closest grapevine, Rod continued. "If you look closely at the base of the vines, you may notice a swelling. This is the location where the scion–a bud or short section of shoot of the fruiting cultivar–was grafted to the rootstock. Depending on how well the two join, the swelling is either invisible or prominent, as is the case here."

PHYLLOXERA

Getting up, Rod continued. "You may already know that grafting began in France to control the devastation being caused by phylloxera–a root louse–during the latter part of the 1800s. It was also the main reason why breeders developed what are now called the French-American hybrids, such as 'de Chaunac,' 'Maréchal Foch,' 'Seyval blanc,' 'Vidal blanc,' etc. The latter varieties often show sufficient resistance to phylloxera to avoid the need for grafting. This leads to considerable saving for growers. Since our own American cultivars are relatively resistant to phylloxera, it is primarily European (*Vitis vinifera*) cultivars that need to be grafted."

"I have a question," interjected Dave. "I've heard that wines used to be better before grafting became necessary. What does the rootstock do to change the flavor of the cultivar it supports?"

"If anything," said Rod, "it may be through the effect of the rootstock on vine growth. Note that this need not be detrimental. Although some critics pontificate wistfully about prephylloxera wines, there are so many factors affecting wine quality that it's silly to imply that rootstock use is the major factor. In addition, those European critics who wax poetic about prephylloxera wines should be careful. They seem to forget that the vines in Chile, and several regions in California and Australia, grow on their own roots. Does this automatically mean that their wines are superior to those from Europe? Some people may think so, but whether it's due to the vines being ungrafted is quite another matter! Almost all vines in Europe are grafted.

"However, the supposed greatness of prephylloxera wines reflects a more general, deep-seated misconception: that technological advances are inherently detrimental to quality. To me, it's the equivalent of saying that the Model T Ford was better than current models. Modern techniques are no more susceptible to abusive misuse than the old."

"Excuse me," said Claude. "I'm so used to hearing about phylloxera, it seems that I know what it is. However, I'm sure I wouldn't recognize the insect if I saw it. What actually does phylloxera do, and how might I recognize it?"

"That depends on the species of grapevine affected. On some species, phylloxera causes leaf galling–swollen areas on the leaves. If I notice any as we walk along, I'll point them out. On other species, the insect attacks the roots. It was the latter expression that ravaged European vineyards. Its influence was so severe that vines died within a few years."

At this point, Klaus called out from where he had kneeled at the base of a vine. "If you want to see phylloxera leaf galling, I have some here on the leaves of a *rupestris* rootstock sucker."

We all went over to have our first look at phylloxera, at least in its leaf-galling stage. There, on the under surface of the leaf, were small yellowish-green hairy balls of tissue. When Klaus cut one open, we could see the small, plump, aphidlike critters inside.

"Do the root galls look similar?" inquired Dave.

"Not really," said Rod. "Also, they are less obvious, at least to the untrained eye. The appearance varies with the part of root system attacked, and with the sensitivity of the plant. Root tips tend to be clubbed (swollen) and bent. Careful scrutiny will show the presence of yellowish, aphidlike insects adhering to root surfaces or in crevices of older roots." More destructive, though, are the large warty growths that can form on older roots."

"Well, returning to grafting," noted Rod, "most commercial grape production is dependent on the use of rootstocks. Although phylloxera was the primary reason for this, additional benefits now include resistance to other root pests, such as roundworms and the viral pathogens they transmit. In addition, some rootstocks limit the damage caused by harsh winters, drought, acidic soils, and salt accumulation. Grapevine vigor can also be regulated to produce better quality fruit by the appropriate choice of a rootstock. The old proverb 'You can't make a silk purse from a sow's ear' certainly applies to wine. The better the fruit, the better the wine, at least potentially."

Appearance of Phylloxera Leaf Galls

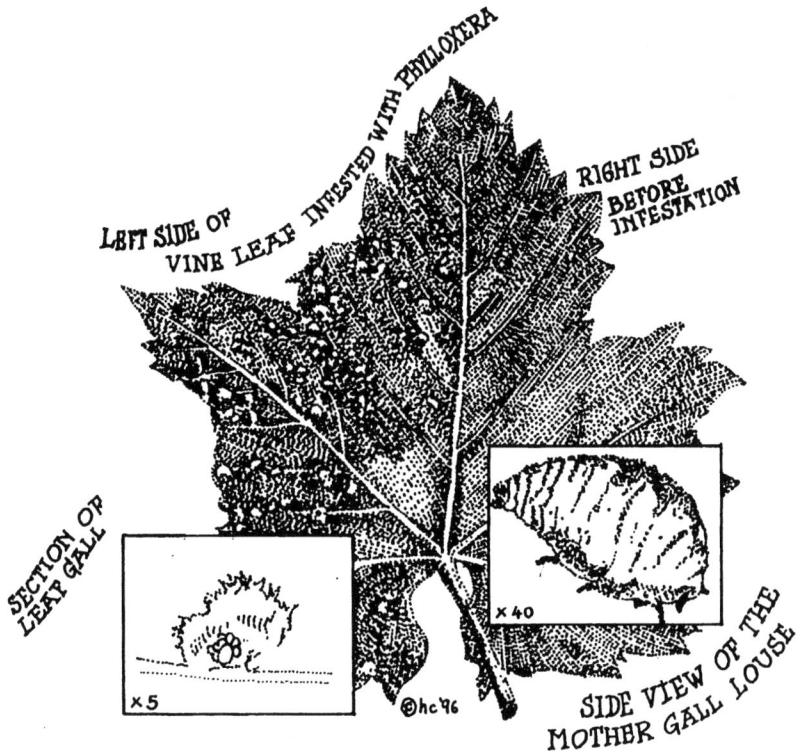

"Another important feature in rootstock selection," added Klaus, "is how local soil conditions affect any particular scion/rootstock combination. As a consequence, we are conducting extensive trials, both here and with local growers, on combinations that appear promising."

"If rootstocks are not of European origin, where do they come from?" inquired Tony.

"They come from here. . . . Well, not exactly New York State, but they are based on North American grapevines," responded Rod.

"North America possesses more grapevine species than any other part of the world. The first rootstocks were all selections from *Vitis riparia* and *V. rupestris*, the riverbank and sand species, respectively. More recently, a wider selection of American species has been used to gain resistance or tolerance to additional problems, such as noted a few moments ago."

"Have I not heard that California has recently had a phylloxera outbreak?" inquired Pat.

"In some regions, yes," responded Rod. "Although a repeat of the devastation that occurred in European vineyards is out of the question, it will be costly to control. Effective pesticides are available, but long-term control will require conversion to different rootstocks. This may require replanting whole vineyards. What may have started the current outbreak was the evolution or accidental introduction of a phylloxera strain capable of attacking the predominantly used rootstock, $A \times R$ #1. Use of several rootstocks probably would have slowed, if not prevented, the present outbreak."

"Getting back to rootstock problems, though," commented Claude, "I remember reading that rootstocks stimulate grapevines into overproducing so much that the vines don't live as long as they used to. Why does this happen?"

"I really wonder where crazy ideas like this come from?" Rod bristled. It was obvious he had heard this view before.

"That growers replant vineyards earlier than they did in the past is one thing; to say that this is due to rootstock use is quite another! Current vineyard replanting cycles are primarily a reflection of today's financial reality. Growers need increased fruit yield for economic survival. Thus, when productivity begins to decline in older vines, the point at which replanting becomes advisable comes sooner. Another factor shortening the productive life of grapevines has been the spread of debilitating viral diseases. Until recently, people didn't realize that supposedly healthy rootstocks could transmit deadly viruses. This rarely occurs nowadays as most commercially available rootstocks are virus free."

"You mentioned vine age a moment ago," commented Phil. "Is there any truth to the view that older vines produce better wine?"

"I really don't know," replied Rod, "though I've heard that view. Since reduced yield favors fruit ripening, the quality of the wine should improve. Have you read anything about this, Klaus?"

"No. If there is any data, it must be obscure," noted Klaus. "I've been reading extensively in preparation for my general doctoral exam and I haven't seen anything."

"Since I have largely monopolized the conversation thus far," noted Rod, "I'll turn things over to Klaus to discuss vine training."

GRAPEVINE STRUCTURE

"Thanks, Rod. However, before discussing training, I'd like to draw your attention to several aspects of grapevines. Notice that some vines, like the one here, are pruned to short stubby segments. These are called spurs. They have two to four shoots arising from each. Other vines, like those to your right, have fewer but longer segments, called canes. Depending on the length, canes may produce more than fifteen shoots. Regardless of whether the vine is spur- or cane-pruned, the number of shoots desired depends on the size and health of the grapevine. This is normally achieved by pruning the vine to retain an appropriate number of buds. Each bud typically produces a single shoot.

"If you look closely, you will notice that some leaves have tendrils arising from the opposite side of the shoot. In wild grapevines, tendrils help the vine climb trees or other supports. Occasionally, wild vines get so large they kill the support tree. Vines remain compact in the vineyards because of pruning and competition for nutrients.

"Two flower clusters usually appear near the base of the shoot, where tendrils otherwise would form. Has everyone found these features?" Klaus asked, looking for anyone who had not found what he was talking about.

"Once you've found a flower cluster, look at the miniature flowers. I have several hand lenses if you need them. Notice how some of the blossoms have opened to show four to five pollen-bearing anthers and a centrally located flasklike ovary. It's from the fertilized ovary that grapes develop. One of the novel features of grapevine flowers is the fusion of the petals at the top! As the blossoms mature, the fused petals separate from the base of the flower and drop away. In the process, pollen is scattered onto the female part of the flower, initiating self-fertilization. Cultivated grapevines are seldom wind–or insect–pollinated.

Basic Grapevine Structure

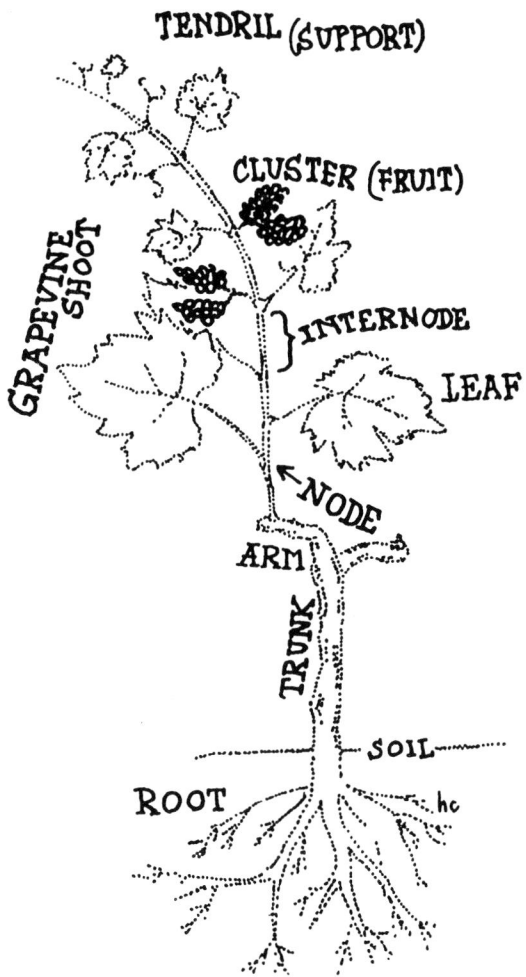

"Just one more fascinating fact about grapevines. Look at the base of each leaf and locate its minuscule bud. Even as we stand here, some of these buds are being stimulated to produce the embryonic flowers for next year. Thus, the conditions around these diminutive buds are setting the vine's future fruiting potential!"

"So each shoot typically produces two fruit clusters?" I said.

"For buds that overwinter, the answer is typically yes. However, a *bud*–called so for simplicity–actually contains four buds. Of these, the outermost (lateral) bud may become active in the year it is formed and produce a shoot. Depending on the grape variety, lateral shoots may flower and produce a second crop. This may seem desirable, but it isn't! Lateral shoot growth and its fruit production retard the ripening of the main crop. It also reduces fruit quality as some of the immature fruit from the second crop is harvested along with the main crop."

Internal Structure of Grapevine Bud

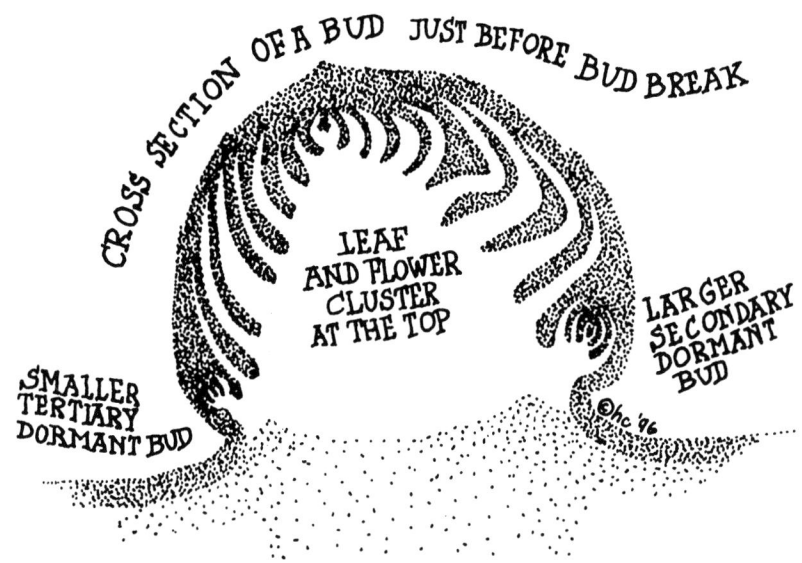

TRAINING SYSTEMS

"Well, if I'm going to talk about training I'd better get to it," reflected Klaus. "Training–how to position the above-ground parts of the vine–is rarely mentioned by wine writers despite its obvious advantages over working on rootstocks." Klaus grinned as he

Development of Grapevine Flower (grape not drawn to same scale)

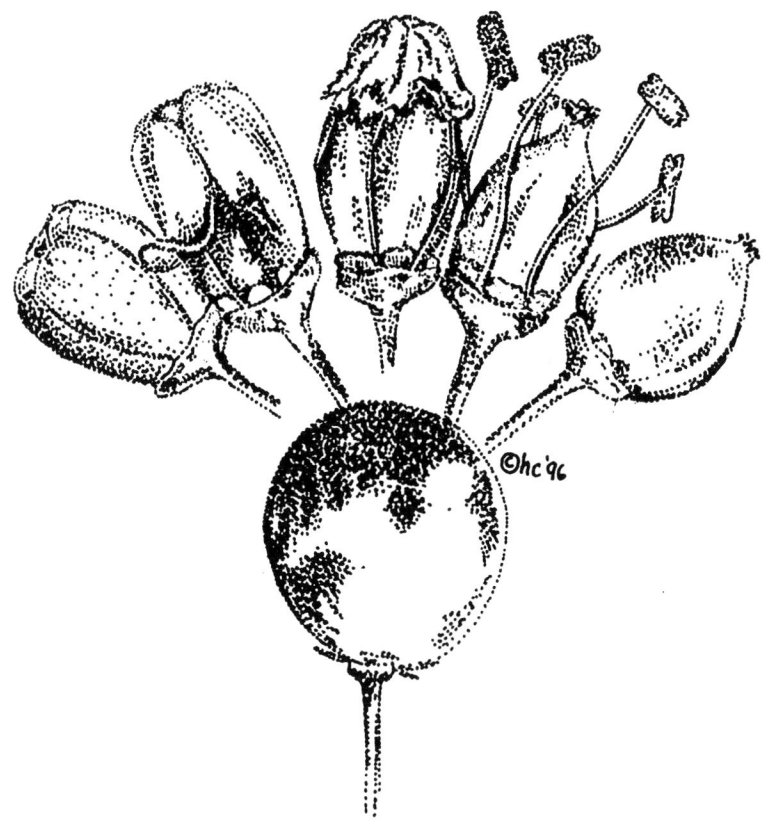

glanced in Rod's direction. "Being above ground, training has a more visible profile.

"Although not a glamorous area of research," Klaus continued, "it is currently very active. Much of the renewed interest was initiated by the pioneering work of Professor Nelson Shaulis, one of the station's former researchers.

"To understand current trends, it's worth remembering how grape growing developed in Europe. Under the subsistence farming economy that existed in Medieval Europe, local food production was crucial to survival. Thus, grape culture was relegated to dry barren sites, or steep slopes unsuitable for food production. These conditions restricted vine growth, but promoted the full ripening of its limited grape yield. Pruning helped to direct vine energy into a few strong shoots.

"In contrast, vineyards were planted on rich lowland soils in the New World. Instead of favoring fruit maturity, severe pruning stimulated excessive shoot growth. This drew nutrients away from the maturing grapes. The luxuriant vegetative growth also produced dense leafy canopies that favored disease development, limited fruit coloration, and retarded aroma production. In response, new training systems direct the energy of vigorous vines toward fruit ripening, and away from excessive vegetative growth.

"One major development has been the separation of the vine canopy (growth) into two or more components. This is especially necessary when the vines are large and planted far apart, as is typical here. The first training system of this type was developed in Geneva, and is appropriately called the Geneva Double Curtain. Training systems more suited to *vinifera* cultivars include the Lyre (France), RT2T (New Zealand), and Scott Henry Trellis (Oregon). Whatever the method, one of the main purposes of canopy division is to maximize sun and air exposure to the fruit."

"How different are these systems?" I asked, intrigued by the canopy-division concept.

"Surprisingly so!" responded Klaus. "Although not too evident as you look at the vines around us now, the Geneva Double Curtain positions the fruit at the top of the vine. This suits the more humid, somewhat overcast conditions we have here in New York. It also reflects the tendency of *labrusca* shoots to trail downward. In contrast, *vinifera* grapevine shoots tend to grow upward. Thus, the upright position of the shoots in the Lyre system is designed for European grapevines. This positions the fruit in a more shaded location, avoiding fruit sunburning in the drier sunnier climates of southern Europe. Other systems try to take into consideration local climatic stresses, the type of wine desired, and fruiting peculiarities of the grape variety. Although fascinating, these details are more than I feel I should get into here.

"However, I would like to mention briefly a radical and novel break with tradition—the minimal pruning system developed in Australia. Minimal pruning, as the name implies, essentially eliminates the yearly removal of most of the shoot growth at the end of the season. Once the basic shape of the vine has been established, the grapevine is allowed to self-regulate its growth. The vine establishes its own balance between root, shoot, and fruit production within the subsequent year or so. Minimal pruning has proven so successful in Australia that we are initiating trials of its efficacy here. The savings in labor alone make it very appealing to growers. Only time will tell whether it will work satisfactorily in the colder, more humid climate here in Upstate New York."

Grapevine Training Systems (Canopy-Division)

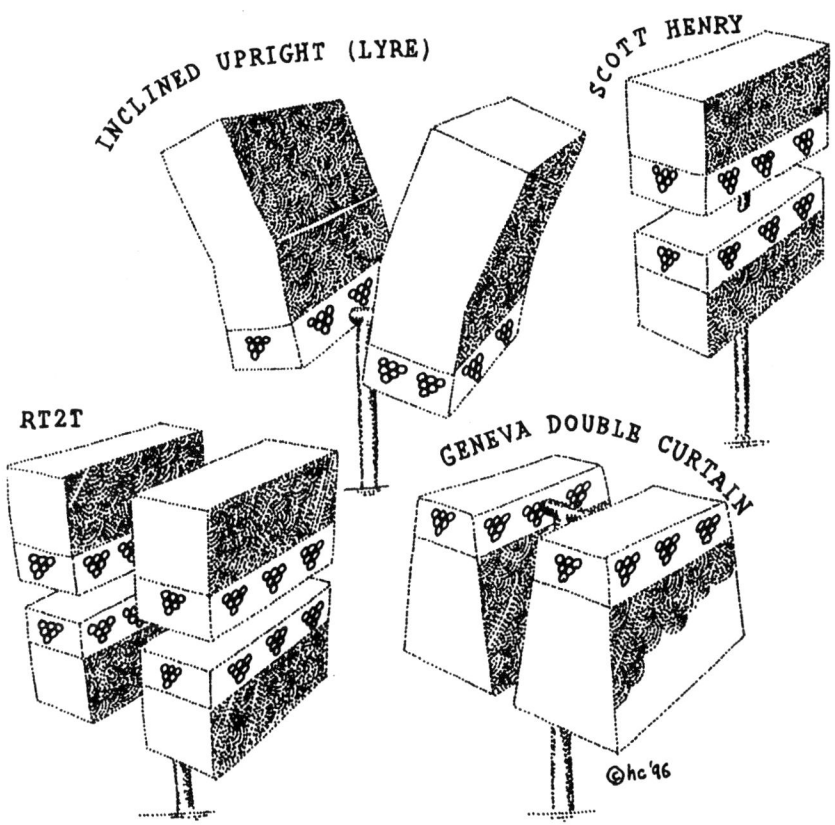

"How can it be that the grapevine can be left on its own, while for centuries pruning was considered essential?" probed Claude.

"The real question is why it took people so long to realize that the vine can regulate its own growth! For millions of years grapevines did quite well without human intervention. Grapevines are unlike many domesticated crops that have been so modified that they can no longer survive on their own. When the vine is left alone, it produces more, but shorter, thinner shoots. Because their ends tend to die and break off during the winter, they do much of their own pruning. Hormonal communication between the root and shoot eventually modulates growth, so that the many smaller fruit clusters mature fully. Excessive shading also tends to be avoided."

"All this sounds great," admitted Dave. "However, I've heard that several well-known producers in California are adopting the old European technique of planting vines close together.[1] The success of the Mondavi/Rothschild venture with 'Opus One' immediately comes to mind. Is not the trend really going the other way–to the adoption of old European techniques rather than your divided canopy systems?"

"I'll admit that the adoption of high-density planting has been getting a lot of positive press recently. I wish our newer systems got a fraction of that attention. Other than its marketing value, high-density planting has two main advantages. Young vines reach peak bearing potential sooner, and vineyards yield more fruit per acre. Because the vines experience greater intervine competition, they tend to remain smaller, generate less self-shading, and bear fully-ripened fruit. This seems great until one realizes the financial cost. Vines, trellising, pruning, and other vineyard expenses can be from three to ten times as expensive as compared with standard vineyards possessing widely spaced vines. In addition, most farm machinery is too wide to navigate the narrow rows of high-density plantings. This means the purchase of new downsized equipment. It's not surprising, then, that wines from such vineyards are expensive. From my experience, the quality improvement, if real and consistent, is not worth the price. Many antiquated techniques survive in Europe only because sufficient numbers of consumers are willing to fork out almost any price for status wines."

---

[1] 10,000 to 12,000 vines per acre versus 2,700 to 3,000 vines per acre.

"So you're all for the newer divided canopy techniques?" continued Dave.

"Insofar as they promote both fruit yield and quality, yes!" responded Klaus. "I may come from a country where tradition is highly regarded, but Germany is also known for its technological innovation. It may seem impossible to increase both yield and quality, but you can! It's possible because severe pruning was developed for inferior growing conditions. With disease-free grapevines–cultivated on rich soils, adequately supplied with water and nutrients, and grown under near ideal climatic conditions–why should the potential of the vine be pruned away? We don't fly jets at the speed of a dirigible. Why, then, train vines in ways that unnecessarily cripple them?"

ORGANIC VITICULTURE

"Are you against organic viticulture then?" said Claude.

"Oh my! There's a loaded question," replied Klaus, glancing at Rod. "There is nothing wrong with the ecological concerns of organic growers. What I object to is the implication that standard wines may be pesticide-laden. This is blatantly false! But, before we go any further, I should make sure that everyone knows what is meant by *organic viticulture*. Organic viticulture refers to the practice of grape culture without the use of manmade fertilizers, pesticides, herbicides, etc."

"If I may, I'll deal with fertilizers and herbicides" interjected Rod. "They're part of my preserve, dealing as they do with the soil and root system.

"The reliance of modern agriculture on fertilizers, herbicides, and pesticides arose out of the need to increase productivity and reduce costs. In this regard, consumers must bear part of the blame, if there is any. Although incomes have risen over the years, the proportion people spend on agricultural produce has declined. With the shift from animal to mechanical horsepower, there was also a reduction in the availability of manure. This, combined with the reduction in the relative cost of electricity, made the production of chemical fertilizers less expensive and more readily available.

"Since they are usually dry and odorless, chemical fertilizers are both easier to store and apply than manures. Regrettably, the overzealous use of fertilizers has led to a depletion in the organic content of many soils. This resulted partially from the nitrate-enhanced bacterial breakdown of humus. The nitrate content of most fertilizers, being water soluble, is also readily leached out of the soil. This in turn has led to contaminated ground water, streams, and lakes. If applied when needed and only as necessary, chemical fertilizers should neither damage the soil nor pollute water supplies."

"But isn't that the problem?" Claude asked. "The chemical industry promotes excessive use to increase sales and profits."

"I will not deny that some in the industry stretch the truth, notably those in advertising," responded Rod. "However, it is a common human failing to think that more is better. Thus, growers may be willing accomplices in overfertilizing. Although industry members may not be perfect, environmentalists are not always faultless either. In their zealous desire to protect the environment, they can act in a manner where they're like 'the pot calling the kettle black.'

NUTRITION AND IRRIGATION

"What we need, though, is not more mud-slinging. What we're trying to do here at the Research Station is to develop earlier and more accurate means of measuring grapevine nutrient needs. The better we are at this, the easier it will be to show that excessive fertilization is economically wasteful and improves neither yield nor quality. This should be more effective in curbing overfertilization than any other factor."

"But, is not animal manure more natural and inherently better?" probed Claude.

"If the manure is well-aged, and isn't diluted with straw, it has several well-known advantages. Aged manure releases nutrients slowly and is less likely to cause water pollution or root burning. Aged manure also adds to the soil's organic content, and improves its textural qualities. However, fresh manure is foul smelling, tends to burn roots, and is a pollutant. In addition, manures may contain toxic accumulations of copper and may induce zinc deficiency in grapevines. Even winery waste can cause manganese toxicity when added to acidic vineyard soils. So, as with anything else, the chemical makeup of the manure must be known for effective use and to avoid creating unforeseen problems.

"Remember that the nutrients taken up by vine roots are identical, whether they come from manure or chemical fertilizers."

"What about herbicides?" interjected Claude.

"Part of the argument against herbicide use is that they damage the soil's structure. This can occur by disrupting the activities of earthworms and other members of the soil fauna. Nevertheless, controlling weed growth with cultivation also damages soil structure. Controlling weed problems by planting ground covers maintains good soil structure and reduces nutrient runoff. However, ground covers are not without their own problems. They can induce nutrient deficiencies by competing with the vine for nutrients, as well as increase frost damage by slowing the rate of heat radiation from the soil. Thus, all weed-control solutions have their drawbacks. Studying the long-term effects of various weed-control measures is one of our mandates at Geneva.

"I hope you have noticed how some rows are tilled, other cleared with herbicides, some mulched, and still others have grass or grass/clover cover crops.

"Ground covers can also benefit biological pest control by harboring predators of grapevine pests. Regrettably, ground covers may also shelter grapevine pests. There are no perfect solutions."

"I often hear people talking about soil structure," remarked Dave. "However, I'm not really clear as to what they're referring. Could you briefly explain what is meant by soil structure?"

"Soil structure refers to one of the soil's more important properties. It denotes how soil particles—such as humus, clay, silt, and sand—form clumps. Structure affects features such as the drainage, water holding capacity, compaction properties, and friability (the ease with which soil clumps crumble). Soil structure also affects the ease with which roots penetrate soil."

Those of us less agriculturally oriented followed Rod's example of illustrating soil friability by picking up clumps of soil and crushing them between their fingers. The rest of us silently enjoyed the sight of city folks getting their hands dirty.

"Well! This seems about as good a time as any for me to pass the discussion over to Klaus. More important, I have to get my son to a baseball game, otherwise my name will be mud."

With that, Rod was off trudging back up the slope before we could thank him for his presentation. His house was just across the road from the research vineyards–very convenient for doing field studies. We had only time to wave goodbye. It was now Klaus' turn to respond to Claude's query about organic viticulture.

"It seems strange for me to talk about vine problems with two plant pathologists here," said Klaus. "I trust you will correct me if I make any mistakes.

PEST AND DISEASE CONTROL

"One of the major features of organic farming is the use of biological pest control. In this aspect, standard viticulture provides one of the best and first examples of the long-term value of biological control–the use of grafting against phylloxera. Rod also noted the use of rootstocks in limiting the damage caused by nematodes (roundworms) and the viruses they transmit.

"Another form of biological control involves the use of natural predators and diseases of grapevine pests. To date, we have had our greatest success against insect pests. Pests often have indigenously occurring predators and parasites. Their effectiveness often depends on modifications that favor the survival of control agents in and around the vineyard. Regrettably, biological control is more finicky and less predictable than chemical applications. Thus, exclusive use of biological control can result in serious crop losses when conditions are favorable for the pest. Therefore, my preference is to have pesticides available, but employed only when needed to avert crop failure. Application of highly selective compounds also tends to minimize its disruption of control agent populations."

"In most garden centers you can now find insecticidal soap mixtures. Are these safer than chemical pesticides?" asked Dave.

"They are safer for humans to be sure; however, they are generally less effective," responded Klaus. "Soaps are also nonselective. That is, they can be as harmful to biological control agents as the pests for which they are used. Note also that insecticidal soaps and oils are no less chemical than manmade pesticides, and not necessarily more *natural*. Furthermore, I know of no research showing that the widespread use of soaps, oils, etc. is harmless to the environment.

"Now, don't get me wrong. We need to reduce our dependence on pesticides of all sorts, not only because insects are so effective in developing resistance to them, but also because they often have unsuspected and undesirable side effects."

"If I could add something?" I said, "One of the 'problems' with chemical sprays was their initial phenomenal success. When growers realized their effectiveness, they became overly reliant on pesticides. As a consequence, older control measures, such as sanitation, often fell into disuse."

"If pesticides are used, are the grapes washed before they are used in making wine?" asked Claude.

"No," replied Klaus. "It's unnecessary. Most pesticides break down while still in the vineyard. In addition, fungicides remaining on the grapes usually precipitate, or are degraded during fermentation. Consequently, pesticides are rarely detected in wine."

"What about breeding resistance into grapevines? I've seen articles extolling the potential of genetic engineering," added Phil.

"Potential, yes," I responded. "Regrettably, genetic engineering is incredibly expensive and the results unpredictable. I can't see investors putting up the millions necessary to insert disease-resistance genes into the multiple range of grape varieties presently cultivated. The financial return simply isn't there."

"What about traditional plant-breeding techniques?" said Phil.

"French plant breeders at the turn of the century tried just that," commented Klaus. "However, the French-American hybrids they developed proved so popular that their planting has been banned in Europe. Well, to be honest, their disease resistance wasn't the reason their planting has been prohibited. Their enhanced yield and reduced production costs made the hybrids very popular–so popular that they began to threaten the traditional flavor of wines produced even in famous wine regions. But, before we condemn Europeans too much for their lack of vision, remember how conservative consumers are. How often do we try wines produced from unknown grape varieties? Consumers tend to be wary of most new products, be they wine, apples, or toothpaste.

"The only place I can see for new, disease-resistant grape varieties is in organic viticulture," continued Klaus. "Consumers interested in organically grown wines tend to be more interested in its organic than varietal nature."

"While you have been discussing the pros and cons of organic viticulture, I have heard no mention of its effect on wine quality. What about that aspect?" questioned Dave.

"The only information I have," commented Klaus, "is a report about off-odors produced by saponified vegetable oils used in controlling insect pests."

"I have tried most of the organically produced wines available in town," said Claude. "My wife and I purchase organically grown produce whenever possible. Nevertheless, I must admit that most of them have been unimpressive. They're also somewhat pricy."

"I've presented several organic wines in class," added Phil. "The best I've found yet is a Fetzer. Maybe when a few more mainstream wineries get into the market, we'll see better tasting organic wines. However, I can't foresee organic wines becoming a major segment of the wine market. Most consumers are simply too price conscious."

"As we've walked through the vineyards, I've noticed that most vines appear to have multiple trunks," noted Tony. "They look so different from pictures of vines I've seen from California. Why are our vines not trained on thick trunks?"

"The use of several separate trunks for each vine reflects the importance of crown gall here in New York," responded Klaus. "The causal bacterium induces gall formation following frost damage during the spring. The gall disrupts water flow in the trunk and favors subsequent infection by rot organisms. When severe, galling can kill the trunk. Because of the frequency of crown gall in New York, vines are commonly trained to several trunks. This way, if one trunk dies, the others sustain vine growth and productivity, while a replacement trunk is established."

"There's one aspect of grape culture we haven't discussed—irrigation," said Pat. "Although seldom an important issue here, does it have a negative influence on grape quality, as some wine writers suggest?"

"Like all good things, if overused, yes," responded Klaus. "However, if used judiciously and only to offset a shortage in rainfall, it is not harmful. In fact, it is essential for grape culture in semiarid regions. Irrigation got a bad name where it was used to increase yield to the point that fruit quality suffered. Water diluted the flavors and sugars in the grapes."

By this point, stomach rumbling was occurring more frequently than questioning. Consequently, Tony wondered if we should take Klaus out for lunch, in appreciation for his arranging the outing. As this met with general approval, Tony asked Klaus if he knew of a pizza place in town that served good wine. Klaus responded in the negative, but volunteered the name of his favorite German restaurant. It also possessed one of the finest wine selections of German, New York, and Californian Rieslings in the state. I preferred that choice as I've never been fond of pizza. About the only good thing I can say about pizza is that it's what we had when I first met Suzanne. But that's a story for another time.

So it was off for, what else, *Wiener schnitzel.* It seemed a bit sumptuous for lunch, but we were in no rush. After all, meandering around in a vineyard for several hours had made us ravenously hungry and thirsty.

Chapter 12

Pleasures of the Table

Our next meeting developed from the realization that we had yet to capitalize on Phil's expertise. It also would be my last chance to attend a meeting before leaving Ithaca.

The problem with sabbaticals is that they end all too soon, and come all too infrequently. No matter how much one enjoys teaching, the confines of formal instruction restrict one's freedom.

I can't really remember who suggested the food and wine theme. The idea seems to have evolved out of our desire to have a meal before we left for field work, for vacation, or, in my case, back to teaching. Because of the culinary nature of the meeting, we were sure our spouses would love to share the experience. Thus, we arranged for them to join us for this special occasion. Phil agreed to plan a meal that would demonstrate the concepts he would discuss.

The meal was arranged to take place at the Big Red Barn on the university campus. The Barn had originally been the stables for the President's residence. Now, it acted as a site for functions that benefitted from a warm rustic atmosphere. It had a fully equipped kitchen, in which Phil's student chefs could prepare his creations. Several of the rooms were intimate enough to provide a cozy environment for refined dining. While he would be holding forth in one of the rooms, his students could prepare and begin the presentation of the savory expressions of his concepts.

It was on a glorious Saturday evening that we congregated at the Big Red Barn. It was warm enough for shorts, but still free of the oppressive humidity that would have made our semiformal attire uncomfortable.

Because most of our spouses had not met each other before, the first order of business was for them to become acquainted. The catalyst of social interaction was champagne. Phil had chosen Mumm's Brut from California and France. These would give us an opportunity to compare similar wines from the same firm, but produced on the western coasts of two adjacent continents.

After a period of convivial exchange, Phil took a break in the conversation to invite us to his presentation.

ORIGIN OF WESTERN VIEWS ON FOOD AND WINE COMBINATION

"Under most situations, choosing a wine should be no more complex than selecting vegetables to go with a meal," Phil began. "In much of Europe, the everyday wine is produced locally. Only on special occasions is a fine wine chosen. Wine is simply the food beverage of choice. In several European countries, wine is still safer to drink than the water, and cheaper than soft drinks.

"It's only connoisseurs for whom choosing the 'right' wine is important. In fact, half the joy in planning a meal can be in selecting the wines to accompany it. For my guests too, I hope, much of the pleasure comes from contemplating the wines I'll serve. Regrettably, food and wine combination is often made out to be so complex and intimidating that it becomes a burden. As a result, people may feel the need to apologize for the wine or avoid serving it altogether to escape potential embarrassment.

"The romance of wine can be real, but stress on this aspect has limited its acceptance as an everyday food beverage. If the only reason for eating out were to have a leisurely romantic experience, fast-food establishments wouldn't be popping up like mushrooms."

"Now that you've brought up the subject of restaurants," commented Pat, "what do you do when the wine list lacks even a trace of originality?"

"When I'm in that situation, I choose the House Wine. It's probably no less inspiring than most of those specified on the wine list, but will be less expensive. Usually I choose the red wine. Although innocuous enough to do no violence to the meal, it may at least have some flavor."

"It's not my intention to malign restaurants for their lack of creativity," continued Phil. "They presumably cater to the desires

"Phil always cooks with wine."

of most of their customers. Many North Americans seem little concerned about relaxing over a meal and savoring the qualities of a fine wine. However, were wine to be minimally taxed here, as in Europe, locally produced wines could be available at the price of soft drinks. Then, wine might take its rightful place as an everyday food beverage. Instead, wine is viewed as a status item, and too many people are willing to allow politicians to tax it excessively.

"But back to choosing that *special* wine. To put it in historical perspective, wine selection is a relatively new phenomenon. European cuisine didn't start to emerge from the quagmire of medieval cooking until the sixteenth century. Grand medieval meals involved repeated combinations of soup, meat, fish, poultry, and sweet dishes. With such a chaotic medley, matching wines with the meal would have been impossible. In addition, poor transportation limited the range of wines available even to the richest nobility. Furthermore, without sulfur dioxide, wine in barrels often turned vinegary by summer.

"Improving economic conditions–associated with exploration, a burgeoning middle class, and industrialization–provided conditions where the demand, production, and transport of wine grew proportionately. These changes favored refinements in eating habits. This culminated in the division of meals into a sequence of soup, fish, salad, meat, cheese, and dessert courses by the nineteenth century. Thus, improved wine availability, and the rediscovery of the benefits of wine aging, occurred along with a rebirth of culinary sophistication. Both features probably encouraged the pairing of wines with the meal."

USES OF WINE IN COOKING

"However, long before people became concerned about matching wine with food, wine became intimately associated with meals in grape-growing regions. This included not only wine's role as the preeminent food beverage, but also in food preparation.

"In food preparation, cooks use wine in several ways. Possibly one of the most ancient is as a marinade. The acidic nature of wine tenderizes meat, temporarily preserving it as well. Wine vinegar has been used even more extensively in pickling foods. Wine also helps to extract or mask the gamy flavor of wild meats. Because the marinade is usually discarded, the wine seldom significantly modifies the food's flavor.

"Another long-established culinary use of wine is in poaching, stewing, or braising. Because cooking dramatically changes the wine's flavor, it is senseless to use expensive or fine wines. The prime concern is that the wine does not adversely affect the food's flavor or appearance. Even the wine's color is seldom of concern, as prolonged heating turns the wine brown."

"OK, so a meal without wine is like a day without sunshine, but you're heading for a sunstroke."

"Does this mean that the use of red burgundy in *coq au vin* is superfluous?" inquired Gayle.

"If you're unconcerned with being traditional, yes," responded Phil. "As far as taste or other aspects, almost any other dry red or white wine would do," answered Phil.

"From what you have just said," commented Mary, Dave's wife, "the use of a champagne in cooking seems unjustified."

"Indeed." Phil nodded. "Carbon dioxide in the wine escapes during cooking, even more rapidly than does its alcohol content.

The only reason for using champagne is for appearances, unless added just before serving.

"When poaching or braising in wine," Phil continued, "the fluid is often reduced to make a sauce. Alternately, wine may be added to deglaze the pan. Because deglazing exposes the wine to less heating, the sauce will possess more of the natural flavors and color of the wine. Remember, the more one wishes the natural flavors of the wine to appear in the food, the later the wine should be added. This also means that selecting the wine requires more care.

"Whereas sweet wines may be paired with dessert, dry wines may be used as a fruit marinade, act as a poaching fluid for firm fruit, be incorporated into a sherbet, or function as a blending medium for creamy custards.

"In its more traditional role as a food beverage, wine plays several functions. At its simplest, wine acts as a palate cleanser. By rinsing away food particles and substituting its own flavor, wine minimizes sensory fatigue. As a result, food maintains its flavor. Conversely, the food helps to freshen the palate for the wine. Wine also may enhance the sensation of certain food flavors, and vice versa. Some food flavors become more soluble and volatile in wine. Equally, some aroma compounds of the wine escape more easily when its alcohol content is diluted in the mouth."

CONCEPT OF FLAVOR BALANCE

"This sounds intriguing, but how do these factors actually enhance flavor perception?" said Klaus.

"The tongue can only sense compounds in solution. This is helped by the acids and alcohol found in wine. Equally, release of aromatic compounds may be promoted by alcohol evaporation.

"Much of what we sense as flavor is actually a combination of taste and smell," continued Phil. "Everyone knows how food loses its flavor when you have a cold. You can reproduce this effect by holding your nose during eating. This retards the diffusion of aromatic compounds from the mouth into the nasal passages. The test also shows how flavor detection requires aromatic compounds to continually pass the olfactory receptors in the nose.

"Wine also plays an important entertainment and status role, especially for connoisseurs. However, wine's most fundamental association with meals is as a flavorant. This is also wine's most complex interaction with food. Its complexity arises from both sensory idiosyncrasies and cultural influences. The physical surroundings and emotional circumstances of a meal also significantly alter our responses to particular food and wine combinations. Why else would regional wines seem so marvelous on a relaxed vacation, but mundane back home with yesterday's lukewarm leftovers. But, to understand accepted norms in food and wine combinations, it is useful to reflect on the nature of cuisine.

"Elizabeth Rosen has made perceptive observations about world cuisines. She has grouped culinary styles based on their use of primary ingredients, cooking techniques, and unique flavorants. Of these, the most distinctive is the use of flavorants. For example, Oriental, East Indian, Mexican, and Italian cooking are respectively characterized by their use of soy sauce, curry, tomato and chili peppers, and a sauce combining olive oil, tomato, garlic, and herbs. These condiments give regional cuisines their character.

"To many outsiders, regional seasoning gives a monotonous similarity. However, the incredible variation in the types of chili peppers, curry preparations, soy sauces, etc., provides a rich diversity of sensory nuances to those habituated to the basic sensation. This is equivalent to the apparent similarity of wines to those unaccustomed to its consumption.

"Especially interesting is the appreciation of the burning sensation of chilies, the bitterness of coffee, or the sourness of pickled foods. The rapid and widespread acceptance of intense flavors, initially perceived as painful or harsh, is in stark contrast to the slow spread of neutral-flavored foods such as corn. This suggests that peer pressure is as powerful a force in shaping food preferences as it is in other aspects of society."

"Interesting as these observations may be, what do they have to do with wine and food combination?" questioned Phil rhetorically.

"I contend they tell volumes! As noted earlier, the pairing of wines with food is little more than three hundred years old. Much of this evolved under the influence of French-modified Italian cuisine, itself adopted from the Near East during the Italian Renaissance.

Thus, the acceptance and appreciation of acidic, often tannic, wine may be accidental. Many Americans, growing up with sweet-tasting beverages, consider dry table wines vinegary. As with chilies and black coffee, only a small proportion of the population, unaccustomed to these tastes early in life, freely adopt them in adulthood. Even here, societal pressure may be central in shaping adult preferences to accepted norms.

"Because food preferences are culturally influenced, value judgements must be viewed relative to the norms on which they are based. For example, sweet/sour combinations are occasionally accepted with both the main dish or dessert. However, acidic wines are rejected with dessert."

INFLUENCE OF FLAVOR PRINCIPLES

"This leads us into what vinous tastes and flavors harmonize with food. Of the four taste sensations we detect, wine possesses only three–sweet, sour and bitter. Since most basic food ingredients exhibit neither sour nor bitter tastes, there is little obvious logic in their association with food. However, food does depress the sourness, bitterness, and astringent aspects found in many table wines. The proteins in food react with the acids and tannins in wine, minimizing their sensory impact. This results in the wine tasting smoother, less sour, and better balanced. Thus, in many cases, it is the food that enhances the perception of the wine, rather than the reverse! Nevertheless, the acidity of wine tends to freshen the mouth, moderate bitterness and astringency while enlivening bland foods.

"Although the taste of most foods is not inimical to wine, several flavorants are, at least to the sensibilities of most Europeans. Vinegar and vinegar-based condiments enhance the sour taste of table wine, making them harsh even to those who relish dry wines. The burning sensation of chilies, and most curries, deaden the taste buds to the subtleties of wine. In addition, heavy doses of spices mask the refined attributes of wine."

"Is this the reason experts suggest that fine wines should be paired with simple dishes?" I asked.

"Exactly!" responded Phil. "Conversely, young flavorful wines are best paired with savory foods. This highlights one of the central concepts of food and wine combination–the matching of flavor intensities so that neither the wine nor the food dominates. Exceptional pairings enhance the mutual appreciation of the food and the wine. Regrettably, it's more difficult to predict the flavor intensity of a wine than food.

"Food flavor intensity and character are often dramatically affected by the use of flavorants and the means of preparation, i.e., poaching versus pan frying. Serving temperature also distinctly affects flavor intensity. Likewise, grape maturity, duration of juice/skin contact, fermentation method, and the form and duration of aging all influence wine flavor. Although most people know how preparation affects food flavor, they are far less aware of how wine gets its character. Furthermore, labels seldom provide useful clues to a wine's flavor. Thus, unless one has tried the wine before, one rarely can accurately predict how a wine will taste.

"Wines and foods seldom have similar flavor qualities. For example, the predominant flavor qualities of wines, such as fruitiness, floral notes, vegetal, and oaky aromas, are rarely found in the basic ingredients of a meal. Conversely, common food flavors are seldom found in wines. Occasionally, though, a flavor component of a wine may complement a similar essence in the food. Examples are the nutty aspect of cream sherries and a walnut dessert, or the oaky character of wines and vanilla cream sauce.

"Other aspects of flavor in matching wines and foods are their complexity, subtlety, balance, and duration.

"Although one typically aims to have equally intense food and wine flavors, this is not always so. A mild wine may be chosen to moderate the strong food flavors. Conversely, a delicate food or cheese may be chosen to highlight the complex subtlety of a well-aged red wine. Nevertheless, some of the most reputed food and wine combinations are apparent opposites. They mysteriously produce compatible interplays of sensations. Examples are the sweet aromatic character of sauternes or port with blue cheeses, or the sharp acidity of a Chablis or Sancerre balanced against the sweetness of crab or the racy aspect of a goat cheese. Even strawberries in a glass of well-aged Bordeaux is considered marvelous. Whether or not these matches are made-in-heaven depends on your subjective reaction to the combination, or one's willingness to accept the authority of gastronomic gurus."

"You have not prescribed the oft-mentioned dictum of red wine with red meats and white wine with fish. However, the expression seems to crystallize much of what you've just said," noted Ruth (Klaus' wife).

"You're right. The 'red with red, white with white' rule focuses attention on balancing savory, dark-colored meats with flavorful red wines, and the milder tasting, pale-colored meats and fish with the delicacy of most white wines. Regrettably, the expression glosses over the marked effects of preparation techniques and condiments on food flavor. The adage also neglects the influence of other important aspects of flavor perception, namely its complexity, development, and duration."

"Traditionally, wines are supposed to go well with cheese, thus the annual circuit of wine and cheese parties. In most instances, though, I don't find that wine and cheese go well together. What are your views?" questioned Pat.

"Like yourself," replied Phil, "I'm less than enthralled by wine and cheese parties. They are rarely good environments in which to assess either the wine or the cheese. But, admittedly, the intent of most such gatherings is not to gauge wine quality.

"Strong-flavored cheeses often mask the subtlety of wines, and mild cheeses contribute little to wine appreciation. Salty cheeses can, though, miraculously reduce the bitterness and astringency of many a tannic red wine. Fine flavored cheeses can also enhance the apparent quality of mediocre wines.

"In the European tradition of grand meals, we shall have a cheese plate just before dessert. During this course I want you to try the Gewürztraminer with Roquefort cheese. You may also wish to try other wine/cheese combinations to see how they enhance, detract, or unaffect the flavors of the cheese and the wine."

"What is your view concerning the use of more than one wine with each course?" inquired Dave.

"In the meal that will commence shortly, you will be experiencing several multiple wine/food combinations. Specifically, we will compare single and pairs of wines with a single course, and two wines simultaneously with two food preparations. Single wines concentrate your attention on the balance, complexity, and subtlety of the interactions between the wine and the food. The use of two wines is more complex, since one compares the interaction between both wines and the food. Even more demanding is the four-way comparison of a pair of wines with two food preparations. Because these comparisons require so much concentration, they are best left to occasions where sensory experience is the prime goal of those around the table.

"Particularly informative are comparisons of varietal wines from different regions. In such, it is important that the identity of the wines should be initially unknown. Too often knowledge of a wine's origin biases one's perception of quality. Conformity to accepted dictates can be so strong that expensive wines remain coveted despite repeated disappointing experiences."

"I hope we're not expected to identify the wines we're trying tonight!" commented Suzanne.

"No!" emphasized Phil. "That type of one-upmanship has nothing to do with wine and food appreciation. Wine enjoyment is no more dependent on your ability to name the wine, than a love of nature requires you to name everything you see."

As Phil glanced toward the door, he announced, "From the looks I'm getting from my students, they're anxious to start. So, if you'll adjourn to the next room, we'll start sensing the realities of food and wine combination. Because this is our group's first dinner, I've had the students print up the menu as a souvenir."

"If I may interrupt for a moment?" requested Dave, "I have a few words before we move. Knowing from experience that Phil would give us a professional presentation and create a memorable culinary experience, I've arranged with the fellows to obtain a suitable liquid memento of our appreciation. On my last trip to New York, I persuaded one of my compatriots to divest himself of a bottle of 1951 Beaulieu Vineyards Cabernet Sauvignon. It is given to you on the condition that it not be opened tonight. The wine is intended for some future occasion."

Phil was visibly shaken by the gift. His hands trembled as he took the bottle. "I'm speechless! For years I've wanted to possess an aged BV Private Reserve produced by Andre Tchelistcheff. Never did I think that I'd possess one from such a great vintage." Looking directly at Dave he said, "Your friend must be especially unique to part with such a wine! I shall place it in an honored position in my cellar. The wine will be held in reserve, awaiting an occasion when we hopefully all meet again and recollect our wonderful year together."

With that, we were off to the adjoining room, from which such delectable aromas had been wafting. We needed no aperitif to get our digestive juices flowing. Sitting down on the burnished cherry-wood Shaker chairs around the oval colonial table, we sighted our menus on ceremonial pewter plates. The menu was a masterful piece of design itself, appearing to be printed on parchment. The menu alone gave ample notice that the evening was never to be forgotten!

"You'll notice that each sitting is provided with numerous wine glasses," Phil commented. "That way, you can keep a little of each wine throughout the meal to follow the wine's development. The courses have been timed to allow you to savor each morsel and assess each drop of wine. Tonight we will dine in the grand manner, and satisfy our souls as well as nourish our bodies. There should also be ample time for discussion."

With a chuckle, Phil commented "Although enjoying the decor and the social aspects of the evening, I find the words of Samuel Johnson provide enlightening council. He's purported to have said:

'This is one of the disadvantages of wine; it makes man mistake words for thoughts."

EXAMPLES OF PRINCIPLES IN ACTION

"As each course is presented, I shall give a brief introduction concerning its preparation, why I chose it, and the wines selected to accompany them.

"The first course is a light almond soup chosen to complement the subtle nutty flavor of our fino sherry, the most delicately flavored of sherries. I chose Tio Pepe because of its popularity. This means that it sells quickly and one can be fairly certain that it will possess its authentic character when purchased. Fino sherries retain their intended flavor only for about six to twelve months after bottling."

CHAMPAGNE
California and French Mumms Brut

SOUFFLÉ OF CULINARY MUSING
By Phil Epstein

ALMOND SOUP
Fino Sherry: Tio Pepe

COQUILLES ST. JACQUES
Chardonnay: Mayacamus & Montrachet

AFRICAN SALAD

BROILED PORK TENDERLOIN
BRAISED PHEASANT BREAST
Rioja: Marqués de Cáceres &
Lopes de Heredia Viña Tondonia Reservas

CHEESE PLATE
Willm Gewürztraminer

DESSERT
Icewine: Vinifera Wine Cellars

PORTS

THE BIG RED BARN

Dinner Menu at the Big Red Barn. (The name "The Big Red Barn" used with permission of Cornell University.)

"This isn't a property of other sherries, is it?" asked Tony. "I thought sherries kept their character almost indefinitely."

"You're correct," replied Phil. "Other types of sherries are more oxidized than fino sherries. As a consequence, they retain their character for months, even after the bottle has been opened."

I had never had almond soup before, nor had I even thought that almonds might be used in a soup. Nevertheless, Phil's choice was astute. The soup did highlight the nutlike flavor of the sherry, and the sherry enhanced the subtle flavor of the soup. From the attention being paid to the pairing, it was clear that others also thought the match was well-conceived.

"While the plates are being cleared away," said Phil, "I'll make a few remarks concerning the Coquilles St. Jacques. The recipe has been modified with the substitution of sole and crab, in lieu of the more usual scallops. The different flavors and textures of each combine to produce an intriguing amalgam designed to harmonize with the complexities of our two Chardonnays. The light Béchamel sauce has been given a dash of vanilla to bring out the oak used in maturing the wines.

"For wines, I've chosen two distinctly different Chardonnays. The Mayacamus Chardonnay from Napa represents the rich, opulent, oaked elegance of what some consider the classic California style. This is compared with the perfumed lusciousness of the Montrachet from southern Burgundy. Both should marry well with the complex essences of the sole and crab. The limpid smoothness of the sauce should harmonize with the mouth-feel of the wines. Which, if either, especially appeals to your taste is important only as a guide to future purchases of similar wines."

There was considerable discussion but no consensus on which of the wines went best with the Coquilles. After we had come to our own conclusions, Phil identified the wines. They had been served in crystal carafes identified by letter only. It was interesting to watch the reactions as the names were announced. Facial expression clearly showed whether one had successfully identified the wines, and whether one considered them worth the price.

After the Coquilles, we were served a tossed green salad. The addition of coarsely chopped peanuts, currants, shredded coconut, orange rind, and sesame seeds gave the salad its African touch. It was a refreshing palate cleanser and set the stage for *la pièce de résistance.*

"For the main course," began Phil, "I have set up a four-way comparison: two wines and two meat dishes. Both wines are from Rioja and are reservas from the same vintage. The predominant grape variety used in the wines is 'Tempranillo.' The Marqués de Cáceres is a comparatively new winery which uses techniques similar to those presently employed in Bordeaux. The wines are aged in 250-litre oak barrels for about two years before being bottled for further aging. Lopez de Heredia is an old firm that retains traditional techniques that harken back to those used in Bordeaux before the 1880s. The wines are aged in large oak tanks for several to many years before bottling. The modern technique, exemplified by Marqués de Cáceres, highlights the varietal fruitiness of the wine, while the older tradition accentuates the development of an aged bouquet and subtle vanilla/oak nuances.

"The two meat dishes are pork tenderloin, roasted over grapevine cuttings, and braised breast of pheasant in truffle sauce. Both were split and filled with stuffing. The different textures and flavors of each dish should provide interesting comparisons with the wines. Both preparations have also been sliced diagonally and placed slightly overlapping on a bed of wild and cultivated rice cooked in chicken stock. Drippings from the pork and deglazings from the braising pan were used to make separate sauces. Truffle slices were added to the pheasant sauce late in deglazing. The plate has been decorated with fried slices of fresh Chinese water chestnut, tree tomato sections, watercress, and cross sections of ripe star fruit. If you have not tried these vegetables before, I think you'll be pleasantly surprised."

I was so involved in sampling the various items, alone and in combination, that I had little time to socialize. Periodically, though, I reflected on the rare opportunity I had had to meet such fascinating people, and to have gained so much in their company.

After the main course came a lemon sorbet. It was new to me in its role as a palate cleanser. My previous experiences with sorbets had been with their sweeter, dessert version.

Next came the cheese plate. It included the Roquefort cheese that Phil specifically wanted us to sample with the Willm Gewürztraminer from Alsace. Also included were samples of Parmigiano Reggiano, Camembert, Cheshire, Gruyère, Brie, and Stracchino, as well as some *fois gras*. These provided a range of flavor intensities and qualities. Although many of the combinations were liked, there was no agreement on which cheese went best with which wine. My personal favorite was not a cheese/wine combination but the Gewürztraminer/*pâté* association.

Finally came the dessert. Thankfully there was only one! Although the meal and the wines had been unsurpassed, it would take a new wardrobe to survive living in such style.

Dessert consisted of a cream custard mousse enveloping thin slivered segments of apricots, peaches, and Bing cherries, topped with whipped cream and passion fruit. This was served with a Vinifera Wine Cellars 'Riesling' Icewine. For the aficionado of dessert wines, this was a heavenly vision transformed into limpid reality.

How long we stayed after the meal sampling ports I can't remember. After such an experience, none of us were anxious to return to our ordinary earthly existence.

DEPARTURE

Welcome or not, the reality of our departure from Ithaca was approaching. It was time for our modular furniture to change back into cardboard containers. Packing time had arrived. Although most of our purchases had been of the consumable variety, we still had accumulated some more tangible items. These included our colonial maple rocker, a folding drafting table, oil painting supplies, two Silverstone frying pans, five ISO wine tasting glasses, Suzanne's Sanderson painting, a Texas Instrument calculator, one Cornell University jacket, and other odds and ends. There were also about a baker's dozen of superb wines.

Our initial plan had been to have the wines transported across the border by Suzanne's relatives. However, their trip had not materialized. To drink the wine I had now would have been infanticide. Being gentle of heart, I couldn't do this. However, the thought of paying an exorbitantly unjust excise tax on the wines turned this usually law-abiding citizen into a smuggler. I struck on the idea of caching my vinous treasures in the bottom of boxes of books. Surely no reasonable customs inspector would be interested in searching through heavy boxes of technical tomes. Thus, they would never realize that the false bottoms concealed wine. The weight of the boxes would not attract attention because books are always heavy. As an additional protection, though, I placed the boxes in the trunk of the station wagon, as far away from prying eyes as possible.

While Suzanne was conducting her final cleaning, I started the job of jostling the sundry collection of boxes and assorted goods into the wagon. Although initially easy, it soon became obvious that not all was going to go as planned! Thus, some of our carefully arranged packing was undone by transferring the contents to bags. Eventually, even this adjustment didn't suffice. I took to dumping the contents into any gap, crevice, or hole I could find.

Somehow, everything got stuffed in, affixed to the luggage rack, or dangled from the back of the wagon. The Torino now had a jaunty upward thrust and the exhaust pipe hovered precariously close to the pavement. Nevertheless, we were ready to bid farewell to our sabbatical abode in paradise. Although there was no smoke, our eyes were wet when it came time to insert the key into the door for the last time. When we had arrived a few months ago, it had been raining outside; now moisture was welling up from the interior. Shortly thereafter, the keys were surrendered and we started up the hill to Uptown Road. Could it be that the sabbatical had been so short? Was it not but yesterday that we had arrived? At such times one especially wants to believe that God closes one door, only to open another.

After several hours on the road, I began to feel somewhat jittery. We were approaching the border. However, I continued to convince myself that we would look squeaky clean. I had some irises from Dr. Randolph's garden, a famous iris breeder and geneticist at Cornell. Before departing, I had obtained the requisite phytosanitary certificate, noting that the plants were disease free. This should demonstrate our integrity, and we would sail through Customs *sans problèmes*.

We pulled into the Customs parking lot on the Canadian side of the Ivy Lea Bridge. We tried to saunter nonchalantly up to the counter, regardless of our inner feelings. I had my phytosanitary certificate clutched in my sweaty right hand, plus letters confirming that I'd been on sabbatical. Everything was dutifully accepted and noted in quadruplicate. It seemed that we soon would be merrily on our way. Then the lady requested to see my green form.

"What green form?" I queried, my brow taking on the appearance of a newly plowed field.

"The green form you filled out when you left the country," came her gentle reply.

"But I didn't stop at Canadian Customs when I left," I responded.

It quickly became clear that in preparing for sabbatical, I had failed to learn about an important document. On leaving the country, I should have filled out a green form noting everything we intended to bring back. That would establish which goods were not subject to taxation. Now was eleven months too late!

The next step was a host of questions related to the goods we'd purchased and were bringing back with us. The primary concern seemed to be whether we had exceeded the permissible maximum, without having to pay excise tax. We had not thought we might be asked such a question. Thus, on the spur of the moment, we could honestly state that we had not exceeded the limit, or at least we thought not. After signing a form to that effect, it seemed we were finally free to go. However, the freedom was to accompany another officer to our vehicle in the parking lot. It was his job to determine if we'd been playing loose with the truth.

His first request was to open one of the side doors. I gingerly opened the door, hoping that not too much would bound out. At the end of packing, gentle persuasion had been required to coax in the last few items. Happily, only one of my hiking boots, a couple of audio cassettes, and a bag of cutlery fell out on the shimmering pavement. He peered in, asked questions faster than we could answer, and finally went around to the back of the wagon. He peered unsuccessfully through the spokes of our two bikes at what was jammed inside. It was at this point that I heard the request that I feared. He wanted me to open the back gate. I stared at him with such a look that he must have thought he'd asked me to squeeze water out of a stone. With hands clenched, as if in prayer, I asked him if it would be acceptable if I just lower the back window. To have opened the back would have required detaching the bicycles and risk a deluge. After a few moments reflection, he accepted my compromise. I darted to the driver's seat to lower the window before he could change his mind. He regarded, almost with disgust, our dilapidated twelve-inch color TV, World War I pots and pans, canary-yellow painted crane-neck lamp, and assorted worthless antiquary. This was not the rich treasure trove that he'd been trained to spy at forty paces. With additional grilling and pokes at visible items, he seemed resigned that we would not provide another notch in his record of impounded scoundrels. So with one final slow inspection of the vehicle, and some scribbling on his pad, he finally acknowledged, almost inaudibly, that we could go.

It was with considerable relief that we began to put back what had been taken or what had fallen out during the inspection. With some extra force, the doors eventually caught, and we could start the car. When I beamed across to Suzanne, indicating my pleasure at escaping, she seemed extraordinarily relieved. This seemed strange as Suzanne is typically super cool. She pointed to my briefcase, squeezed next to the door the customs official had asked me to open. Suzanne stated that it contained the ledger in which we had noted, to the penny, every purchase. It would have chronicled in minute detail that we had indeed exceeded our limit. Alone, the painting Suzanne had purchased in October would have exceeded her limit for the year. Anyway, the ordeal was over–or so we thought.

We had not traveled more than about a thousand yards beyond the custom's parking lot when it happened. A horrific clanging enveloped the vehicle. Instinctively my eyes deflected up to the rear view mirror. That was of no avail. All one could see was a jumble of boxes. Glancing out the side view mirrors revealed no more about the origin of the earsplitting clamor. Could it be the Mounties swooping down in attack helicopters? Visions of jail bars and reports of the prof impounded by the police flashed through my brain. Then a feeling of peace descended. I looked at Suzanne, whose eyes were as large as saucers, and began to laugh.

"It's the ruddy alarm clock!"

For some crazy reason the buzzer had gone off. It was buried in my briefcase. The alarm finally ceased ringing as we turned right onto Highway 401. With a final quick glance back toward the bridge, the last view of New York State disappeared behind the trees.

Sabbatical was over. Happily, though, there was still lots of great drinking ahead!

Bibliography

GENERAL WINE BOOKS

Adams, L. 1991. *The Commonsense Book of Wine*. Wine Appreciation Guild, San Francisco, CA.

Amerine, M.A. and Singleton, V.L. 1977. *Wine–An Introduction*. University of California Press, Berkeley, CA.

Baldy, M. 1993. *The University Wine Course*. Wine Appreciation Guild, San Francisco, CA.

Johnson, H. and Halliday, J. 1992. *The Vintner's Art: How Great Wines Are Made*. Simon & Schuster, New York, NY.

Kramer, M. 1989. *Making Sense of Wine*. William Morrison & Co., New York, NY. pp. 224

Meyer, J. 1989. *Plain Talk About Fine Wine*. Capra Press, Santa Barbara, CA.

Robinson, J. (ed.) 1994. *Oxford Companion to Wine*. Oxford University Press, Oxford, England.

WINE HISTORY

Allen, H.W. 1961. *A History of Wine*. Faber and Faber, London, England.

Badler, V.R., McGovern, P.E., and Michel, R.H. 1990. Drink and be merry! Infrared spectroscopy and ancient Near Eastern wine. *MASCA Res. Pap. Sci. Archaeol.* 7: 25-36.

Henderson, A. 1824. *The History of Ancient and Modern Wines*. Baldwin, Craddock & Joy, London, England.

Hyams, E. 1965. *Dionysus: A Social History of the Wine Vine*. Thames & Hudson, London, England.

Johnson, H. 1989. *Vintage: The Story of Wine*. Simon & Schuster, New York, NY.

Levadoux, L. 1956. Les populations sauvages et cultivées de *Vitis vinifera* L. *Ann. Amélioration Plantes. Séries B*. 1: 59-118.

McGovern, P. and Fleming, S. 1996. *The Origins and Ancient History of Wine*. Gordon and Breach Publishers, Amsterdam, Holland.

Olmo, H.P. 1976. Grapes: *Vitis, Muscadinia* (Vitaceae). pp. 294-298. In: *Evolution of Crop Plants*. Simmonds, M.W. ed., Longman, London, England.

Redding, C. 1851. *A History and Description of Modern Wines,* 3rd ed., Henry G. Bohn, London, England.

Unwin, T. 1991. *Wine and the Vine: An Historical Geography of Viticulture and the Wine Trade*. Routledge, London, England.

Younger, W. 1966. *Gods, Men and Wine*. George Rainbird, London, England.

GRAPE GROWING

Champagnol, F. 1984. *Eléments de Physiologie Végetale et de Viticulture Genérale*. Published by Author.

Coombe, B. and Dry, P. 1992. *Viticulture. Vol. 1 and 2*. Winetitles, Adelaide, Australia.

Hillebrand, W., Lott, H., and Pfaff, F. 1984. *Taschenbuch der Rebsorten*. Facchverlag Dr. Fraund GMBH, Wiesbaden, Germany.

Huglin, P. 1986. *Biologie et Ecologie de la Vigne*. Payot Lausanne Technique of Documentation, Paris, France.

Jackson, R.S. 1994. *Wine Science: Principles and Applications*. Academic Press, San Diego, CA.

Morton, L.T. 1985. *Winegrowing in Eastern America*. Cornell University Press, Ithaca, N.Y.

Mullins, M.G., Bouquet, A., and Williams, L.E. 1992. *Biology of the Grapevine*. Cambridge University Press, Cambridge, England.

Pongrácz, D.P. 1978. *Practical Viticulture*. David Philip, Cape Town, South Africa.

Ribéreau-Gayon, J. and Peynaud, E. 1971. *Traité d'Ampélologie*. Sciences et Techniques du Vin–Tome 1 and 2. Dunod, Paris, France.

Smart, R.E. and Robinson, M. 1991. *Sunlight into Wine: A Handbook for Winegrape Canopy Management.* Winetitles, Adelaide. Australia.

Winkler, A.J., Cook, J.A., Kliewer, W.M., and Lider, L.A. 1974. *General Viticulture.* University of California Press, Berkeley, CA.

WINE CHEMISTRY

Jackson, R.S. 1994. *Wine Science: Principles and Applications.* Academic Press, San Diego, CA.

Liskens, H.F. and Jackson, J.F. 1988. *Wine analysis. Modern Methods of Plant Analysis.* New Ser. Vol. 6, Springer Verlag, Berlin, Germany.

Ough, C.S. and Amerine, M.A. 1988. *Methods for Analysis of Musts and Wines.* Wiley, New York, NY.

Usseglio-Tomasset, L. 1989. *Chemie Oenologique.* Technique et Documentation–Lavosier, Paris, France.

Würdig, G. and Woller, R. 1989. *Chemie des Weines.* Handbuch der Lebensmitteltechnologie, 2nd ed., Ulmer, Stuttgart, Germany.

Zoecklein, B.W., Fugelsand, K.C., and Gump, B.H. 1990. *Production Wine Analysis.* AVI Publishing Co., Westport, CT.

WINEMAKING

Amerine, M.A. and Joslyn, M.A. 1970. *Table Wines–The Technology of their Production,* 2nd. ed. University of California Press, Berkley, CA.

Amerine, M.A. and Singleton, V.L. 1977. *Wine–An Introduction.* University of California Press, Berkley, CA.

Boulton, R.B., Singleton, V.L., Bisson, L.F., and Kunkee, R.E. 1995. *The Principles and Practices of Winemaking.* Chapman Hall, New York, NY.

Dittrich, H.H. 1987. *Mikrobiologie des Weines.* Handbuch der Lebensmitteltechnologie, 2nd ed., Ulmer, Stuttgart, Germany.

Farkas, J. 1988. *Technology and Biochemistry of Wine,* Vols. 1 and 2. Gordon and Breach Science Publishers, New York, NY.

Fleet, G.H. (ed.) 1992. *Wine Microbiology and Biotechnology.* Harwood Academic Publishers, New York, NY.

Jackisch, P. 1985. *Modern Winemaking.* Cornell University Press, Ithaca, NY.

Jackson, R.S. 1994. *Wine Science: Principles and Applications.* Academic Press, San Diego, CA.

Margalit, Y. 1992. *Winery Technology and Operations: A Handbook for Small Wineries.* The Wine Appreciation Guild, San Francisco, CA.

Ough, C.S. 1991. *Winemaking Basics.* Food Products Press, Binghamton, NY.

Peynaud, E. 1984. *Knowing and Making Wine.* John Wiley & Sons, Inc., New York, NY.

Rankine, B.C. 1989. *Making Good Wine: A Manual of Winemaking Practice for Australia and New Zealand.* Macmillan, Melbourne, Australia.

Ribéreau-Gayon, J., Peynaud, E. Ribéreau-Gayon, P., and Sudraud, P. 1972. *Traité d'Oenologie.* Sciences et Techniques du Vin–Tome I-IV. Dunod, Paris, France.

Troost, R. 1988. *Technologie des Weines.* Handbuch der Lebensmitteltechnologie, 2nd ed., Ulmer, Stuttgart, Germany.

Vine, R.P. 1981. *Commercial Winemaking.* AVI Publishing Co., Westport, CT.

SPECIAL WINEMAKING TECHNIQUES

Botrytized Wines

Ribéreau-Gayon, P. 1988. *Botrytis*: Advantages and disadvantages for producing quality wines. pp. 319-323. In: *Proc. 2nd Internat. Symp. Cool Climate Vitic. Oenol.*, Auckland, New Zealand, Jan. 11-15, 1988. Smart, R.E., Thornton, S.B., Rodriguez, S.B., and Young, J.E. (eds.), N.Z. Soc. Vitic. Oenol., Auckland, New Zealand.

Carbonic Maceration

Flanzy, C., Flanzy, M., and Bernard, P. 1987. *La Vinification par Macération Carbonique.* Institute National de la Recherche Agronomique, Paris, France.

Fortified Wines

Goswell, R.W. 1986. Microbiology of fortified wines. *Dev. Food Microbiol.* 2: 1-20.

Goswell, R.W. and Kunkee, R.E. 1977. Fortified wines. pp. 477-535. In: *Economic Microbiology,* Vol I–Alcoholic Beverages. A.H. Rose, (ed.) Academic Press, New York, NY.

Joslyn, M.A. and Amerine, M.A. 1964. *Dessert, Appetizer and Related Flavored Wines.* Univ. of California Press, Berkeley, CA.

Recioto Wines

Usseglio-Tomasset, L., Bosia, P.D., Delfini, C., and Ciolfi, G. 1980. I vini Recioto e Amarone della Valpolicella. *Vini d'Italia* 22: 85-97.

Sparkling Wines

Markides, A.J. 1987. The microbiology of methode champenoise. pp. 232-236. In: *Proc. 6th Austral. Wine Industry Tech. Conf.* Lee, T. (ed.) Australian Industrial Publishers, Adelaide, Australia.

Moulin, J.P. 1987. Champagne: The method of production and the origin of the quality of this French wine. pp. 218-223. In: *Proc. 6th Austral. Wine Industry Tech. Conf.* Lee, T. (ed.) Australian Industrial Publishers, Adelaide, Australia.

GRAPE GROWING AND WINEMAKING

Jackson, D. and Schuster, D. 1995. *The Production of Grapes and Wine in Cool Climates,* 3rd ed. Winetitles, Adelaide, Australia.

Jackson, R.S. 1994. *Wine Science: Principles and Applications.* Academic Press, San Diego, CA.

Morton, L.T. 1985. *Winegrowing in Eastern America.* Cornell University Press, Ithaca, NY.

Weaver, R.J. 1976. *Grape Growing.* Wiley, New York, NY.

WINE GEOGRAPHY

Baxevanis, J.J. 1992. *The Wine Regions of America: Geographical Reflections and Appraisals.* Vinifera Wine Growers Association, Stroudsburg, PA.

de Blij, H.J. 1983. *Wine: A Geographic Appreciation.* Rowman & Allanheld, Totowa, NJ.

de Blij, H.J. 1985. *Wine Regions of the Southern Hemisphere.* Rowman and Allanheld, Totowa, NJ.

Gladstone, J. 1992. *Viticulture and Environment.* Winetitles, Adelaide, Australia.

Johnson, H. 1994. *The World Atlas of Wine.* Simon & Schuster, New York, NY.

Lichine, A. 1987. *New Encyclopedia of Wines and Spirits.* Alfred A. Knopf, New York, NY.

Saayman, D. 1977. The effect of soil and climate on wine quality. pp. 197-208. In: *Int. Symp. Quality Vintage.* Oenol. Vitic. Res. Inst., Stellenbosch, South Africa.

Seguin, G. 1986. "Terroirs" and pedology of wine growing. *Experientia* 42: 861-873.

WINE AND HEALTH

Bisson, L.F., Butzke, C.E., and Ebeler, S.E. 1995. The role of moderate ethanol consumption in health and human nutrition. *Am. J. Enol. Vitic.* 46: 449-462.

Ford, G. 1993. *The French Paradox: Drinking for Your Health.* Wine Appreciation Guild, San Francisco, CA.

Janero, D.R. 1995. Ischemic heart disease and antioxidants: Mechanistic aspects of oxidative injury and its prevention. *Crit. Rev. Food Sci. Nutrit.* 35: 65-82.

Katz, H.J. (ed) 1981. *Proceedings of Wine, Health and Society—A Symposium.* GRT Books, Oakland, CA.

Kaufman, H.S. 1992. The red wine headache and prostaglandin synthetase inhibitors: A blind controlled study. *J. Wine Res.* 3: 43-46.

Lucia, S.P. 1963. *A History of Wine as Therapy.* Lippincott Co., Philadelphia, PA.

Taylor, S.L., Higley, N.A., and Bush, R.K. 1986. Sulfites in foods: Uses, analytical methods, residues, fate, exposure assessment, metabolism, toxicity, and hypersensitivity. *Adv. Food Res.* 30: 1-75.

SENSORY ANALYSIS

Amerine, M.A. and Roessler, E.B. 1983. *Wines–Their Sensory Evaluation*. Freeman Co., New York, NY.

Jackson, R.S. 1994. *Wine Science: Principles and Applications*. Academic Press, San Diego, CA.

McCloskey, L.P., Sylvan, M., and Arrhenius, M. 1996. Using quality experts in descriptive analysis of regional typicalness of Chardonnay wine aroma. *J. Sensory Stud.*, 11:1.

Peynaud, E. 1983. *The Taste of Wine*. MacDonald and Co., London, England.

Stone, J. and Sidel, J.L. 1993. *Sensory Evaluation Practices,* 2nd ed. Academic Press, San Diego, CA.

WINE AND FOOD

Edsrud, B. 1984. *Wine with Food*. Congdon & Weed, Inc., New York, NY.

Galford, E. (ed.). 1983. *Wine*. The Good Cook Series. Time-Life Books, Alexandria, VA.

McGee, H. 1984. *On Food and Cooking: The Science and Lore of the Kitchen*. Scribners, New York, NY.

Rietz, C.A. 1961. *A Guide to the Selection, Combination and Cooking of Food,* Vol. 1 and 2. AVI Publishing Co. Inc., Westport, CT.

Rozen, E. 1982. The structure of cuisine, pp. 189-203. In: *The Psychobiology of Human Food Selection*. Barker, L.M. (ed.). AVI Publishing Co., Westport, CT.

Tannahill, R. 1973. *Food in History.* Stein & Day Publishing, New York, NY.

CONSUMER WINE MAGAZINES

The Arbor Magazine, P.O. Box 20254, Atlanta, GA.

Decanter Magazine, Priory House, 8 Battersea Park Road, London, SW8 4BG, England.

Harpers Wine and Spirit Magazine, Harling House, 47-51 Great Suffolk St., London SE1, England.

Italian Wine and Spirits, One World Trade Center, Suite 2253, New York, NY.

Society of the Medical Friends of Wine Bulletin, Society of the Medical Friends of Wine, Box 218, Sausalito, CA.

Wine Country Magazine, 985 Lincoln Ave., Benicia, CA.

The Wine Spectator, M. Shanken Communications, Inc., 387 Park Ave. S., New York, NY.

Wine and Spirits Magazine, Evro Publ. C. Ltd., 60 Waldergrave Rd., Teddington, Middlesex, TW11 8LG, England.

Wine World, 15101 Keswick St., Van Nuys, CA.

TECHNICAL AND RELATED TRADE JOURNALS

American Wine Society Journal, 3006 Latta Road, Rochester, NY, 14612.

American Journal of Enology and Viticulture, P.O. Box 700, Locke-ford, CA, 95237-0700.

Australian Grapegrower and Winemaker, Ryan Publ., 95 Currie St., Adelaide, South Australia, Australia.

Australian Wine Industry Journal, Australian Industrial Publishers Pty Ltd., 2 Wilford Ave., Underdale, SA, 5032, Australia.

Australian Wine Research Institute Technical Review, Private Mail Bag, Glen Osmond, SA, 5064, Australia.

Bulletin de l'Office International de la Vigne et du Vin, 11 rue Roauépine, F-75008, Paris, France.

Die Wein-Wissenschaft, Fachverlag Dr. Fraund Gmbh, An der Brunnenstube 33-35, Mainz 25, D-6500, Germany.

Die Weinwirtshaft, P.O. Box 312, Neustadt and der Weinstrasse, D-6730, Germany.

Enoteca Italiana, Città del Vion e le Enoteche, viale Partigiani 21, Asti, I-14100, Italy.

Journal of Small Fruit and Viticulture, Haworth Press Inc., 10 Alice Street, Binghamton, NY, 18904.

Journal of Wine Research, Carfax Publ. Co., P.O. Box 25, Abingdon, Oxfordshire, OX14 3UE, England.

Journal Internationale de la Vigne et du Vin, Vigne et Vin Publications Internationales, Bordeaux Technopolis, Site Montesquieu, F-33651, Martillac, France.

La Journée Vinicole, C.P. 1064, 7 rue Dom-Vaissette, Montpellier, F-34007, France.

Mitteilungen Klosterneuburg, Höhere Bundeslehr-und Versuchsanstalt für Wein-und Obstbau mit Institut für Bienenkunde, Wiener Strasse 74, A-3400 Klosterneuburg, Austria.

Practical Winery and Vineyard, Don Neel and Assoc., 15 Grande Pasco, San Rafael, CA, 94903.

Progrés Agricole et Viticole, Société Agricole et Viticole, 1 bis Rue de Verdun, Montpellier, F-3400, France.

Rebe und Wein, Verlag Schwabstr., 20 Postfach 1180, Weinsberg, D-7102, Germany.

Revue Suisse de Viticulture Arboriculture Horticulture, M. Girardin, case 190, Nyon, CH-1260, Switzerland.

South African Journal of Enology and Viticulture, P.O. Box 2092, Dennesig, Stellenbosch 7600, South Africa.

The Vinifera Wine Growers Journal, Vinifera Wine Growers Assoc., 1947 Hillside Dr., Stroudsburg, PA, 18360.

Vineyard and Wine Management, Vineyard and Winery Services Inc., 103 Third Street, Box 231, Watkins Glen, NY, 14891.

Vigneron Champenois, 5, rue Henri-Martin, B.P. 135, Epernay, F-51204, France.

VigneVini, Edagricole S.p.A., via Emilia Lavante, 31, Bologna, I-40139, Italy.

Vini d'Italia, Edizioni A.E.B. spa, Via V. Arici, 92, San Polo, Brescia, I-25010, Italy.

Viticultura Enologia Profesional, AgroLatino, Apdo. 20151, Barcelona, E-08080, Spain.

Vitis, Bundesforschungsanstalt für Züchtungsforschung im Wein-und Gartenbau, Institut für Rebenzüchtung Geilweilerhof, Siebeldingen, D-6741. Germany.

Weinberg u Keller, Verlag, Postfach 1148, Bernkastel-Kues, D-5550, Germany.

Wines and Vines, 1800 Lincoln Ave., San Rafael, CA, 94901-1296.

Wynboer, P.O. Box 528, Suider, Paarl, 7624, South Africa.

Index

Page numbers followed by the letter "t" indicate tables; those followed by "i" indicate illustrations.

Acetaldehyde, 112
Acetaminophen 142
Acetic acid, 64. *See also* Vinegary
Acetic acid bacteria, spoilage
    organism, 41,60
Acetobacter, 64
Acetylsalicylic acid (ASA), 136
Acid
  malic, 15
  tartaric, 15
Acidity, 14,15,25
  aging, 34
  taste, 14,15,25,34
  value of, 15
  wine color, 15
  wine taste, 106
Additives
  chemical, 130
  foreign, 129
Adelsheimer Pinot noir, 22
Adrenalin, 135
Adulteration, 128,130
  definition of, 131
Aeration, 52
Aged wine, value of, 122-123
Ageusia, 117
Aging, 121-125,126i,127
  affected by
    acidity, 34,127
    humidity, 125
    light, 125
    temperature, 124-125
    vibration, 125

Aging *(continued)*
  effect on
    color, 123
    fragrance, 123,126i
Aging potential
  cork length, 88-89
  grape varieties, 90
  indicators, 88-89,127
  national perspectives on, 91
  red wines, 90
  white wines, 90
Aging *sur lies*, 176
Air-bladder, 62
Alcohol, 67,127
  opposition to drinking, 147
Alcohol content
  burning sensation, 108
  weight or body, 108
  wine tears, 108
Aleatico, aroma, 102t
Allergies, food and beverage, 134
Amelioration, definition of, 131
*American Journal of Enology
  and Viticulture,* 28
Amphora(s), 171,171i
  pitched, 171-172
Anosmia, 117
Anthers of grapevine, 192
Anthocyanin, 15,35,157,176. *See
  also* Red wine pigment
AP number, 177
Appellation control laws (AOC), 8-9,
  83-85
  advantages, 83,84

Appellation control laws (AOC)
  (continued)
  and taste tests, 83-84
  barrel use, 174
  disadvantages, 83,84
  in Italy, 84
  significance in marketing, 9
Appelation control system,
     8-9,19,118
Appellations, California wines, 118
Aroma(s), 20,21,67,100
  characteristics, 19-23
  cultural influence, 105
  definition of, 20
  divergent opinions, 105
  French expression of, 105
  German expression of, 105
  Gewürztraminer, litchi nut, 103
  Italian wines, 120
  northern Italian expression of, 105
  sherry, walnut, 103
Aromatic compounds, 100
Arteriosclerosis, 137-138
Aspiration of wine, 109
Asti Spumante, 81,157
Astringency
  vs. bitterness, 109
  mouthfeel, 20
  produced by grapes, 67
Aurora, 157
Auslese, 178
Australian wines, 19,145
AxR#1, rootstock, 191

Babera, aroma, 102t
Bacchus Society, 6,7,12,22,
     27,93,118
Bag-in-box wine, 90
Bailey, L. H., 17
Baked aroma, 151
Baked flavor, 112
Balance, quality of fragrance, 103
Barolos, 9,120

Barrel(s), 173
  during wine maturation, 69
  fermentation in, 175-176
  heat fixing, 174
  length of use, 174
  substitutes for, 175
  toasting of, 173
Barrel making, traditional, 173-174
Beaujolais, 37-38,48,126i
  cru, 48
  nouveau, 48
     headaches, 135
Beaulieu Vineyards Cabernet
     Sauvignon, 217
Beerenauslese (BA), 178
Big Red Barn, 207
Bird repellants, 31-32
Bitterness
  vs. astringency, 109
  detected in mouth, 106
  duration, 106
  of wine, 106
  produced by grapes, 67
Bleach, 170
Blending, of wines, 20
Boiling, 69
Botrytis, 2,24,47,182,184,185
  cinerea, 178,181
  convoluta, 187
  disease of, 2
  fragrance, 179,181
  onion leaf spot, 7
Botrytized wine, 2,19,180,182
  aroma, 101t
  flavor of, 180-181
Bouquet, 20,21
  definition of, 20
  produced by yeast, 67
Brandon University, 5,7,8,47
Brandy, *marc*, 50
Breathing, 120
Brettanomyces, spoilage yeast,
     41,60,64
Brix, 47,48,50

Browning, 123
  oxidative, 61
Brut (rough), 155,156
Buds, grapevine, 194,194i
Bulk process, sparkling wines, 159-160
Bureau of Alcohol, Tobacco, and Firearms, 71
Burgundy, wine, 22

Cabernet Franc, aroma, 102t
Cabernet Sauvignon, 87,90
  aroma, 102t
  grape variety, 17-18,20-21
California wine appellations, 118
Canada, liquor control stores, 8-9
Canadian wine, 116
Cane, sugar, 129
Canopy, 196
Cap, 46
Carbon dioxide, 48,99
  from fermentation, 51
  loss from wine, 51
Carbonic maceration, 36-40,48
Carboys, 38,40,41,43,49,50,51,52
Catawba grape, 17,157
Caucasian region, origin of wine, 13-14
Cayuga blanc, 98
Cayuga grapes, and chaptalization, 66
Centrifuge, 68
Champagne, 88,155,162-163. *See also* Sparkling wine
  history of, 156
  labels, terminology, 156-158
Champagne, region of France, 157
Chaptalization, 66
  and hybrid grapes, 66
  technique, 66-67
  and *vinifera* grapes, 66
Chardonnay, 90
  aroma, 101t
  California, 87

Chardonnay *(continued)*
  grape variety, 17-18,19,157
  at Heron Hill, 71
Château Beychevelle, 10
Cheese cloth, 46
Chenin blanc, aroma, 101t
Chianti(s), 72,120,183
Chlorine contamination, 170
Chlorogenic acid, 128
Cholesterol, 138
Cigar-box bouquet, 20
Cigar-box odor, 124
Cinque Castelli Spanna, 10
Clever Hans Bakery, 8
Climate Near the Ground, 32
Closures, aluminum roll-on, 172
Complexity, defined, 103
Complexity, quality of fragrance, 103
Concord grape(s), 17,157
  and chaptalization, 66
Confrérie de la Barrique, 37
Continuous press, 62
Cork
  adding paraffin or silicone, 169
  ancient usage, 171
  champagne, 160,161i
    opening, 161
  closures, 161,166
  contaminants, 169,170
  growth rings, 166-167
  lengths of, 88-89
  in port, 162
  punching out, 167
  quality, 168-169
    defined, 89
  reproduction, 166
  second, 166
  in sherries, 162
  smelling the, 172
  structure and properties, 166-169
  virgin, 166
Cork-borne faults, 169-172
Cork Oak, bark, 166
Corked, 111,112

Corkscrews, 96-97,98i
Corky, 111
Corky odor, 170,171
Cornell Medical School, 134
Cornell University, 3,7,47
  Hotel School, beverage course, 6
  Plant Pathology Department, 3,6,7
  wine appreciation course, 93
Corvina, 184
  aroma, 102t
Cosimo Taurino's Notarpanaro, 23
Criadera, 154
Crown gall, 204
Cru classé system, 84
Cryoprotectants, 28
Crystal, 94
Cuvée (blend), 158,159
CWDW Society, 11-26
  origin of, 3,6,7,8
  procedure, 11

Damascenone, 124
De Chaunac, 39i,45,46,49,188
Decanter, 133
Decanting, 47-54,120
  scope of, 52
  uses of, 120
Delaware, grape variety, 17,157
Demi-sec (half dry), 156
Dessert wine, 23-24,91
Diacetyl, 12
Divided canopy system, 196
Dolcetto, aroma, 102t
Dopamine, 135
Double-blind study, 134-135
Doux (soft), 156
Drinking, reticence, 116
Drouhin Estate Pinot noir, 22
Duchess, grape variety, 17
DuPont Surfactant WK, 28
Duration
  definition of, 104
  quality of fragrance, 103-104

Egypt, evidence of early wine
  making, 16
Eiswein, 131,178,182
Erbacher Sandgrup Riesling
  Auslese, 176
Esters, 65,68,123,124
Ethyl acetate, 64,113

Fermentation, 38,41,43,46,49,50
  in-barrel, 38
  stuck, 60
  temperature
    control, 43,64
    for red wines, 65
    for white wine, 65
Fermentor, 40,52,62-65
  Heron Hill, 64
  operation of, 64
Fertilization, 31
  flower pollination, 16-17
  vine nutrition and, 31
Fertilizers, 199,200
Fetal alcohol syndrome (FAS),
  143-144
Fetzer, 204
Fining, 140
Finish, 19,20
  definition of, 109
Fino sherry, 151,152,154
Firing, 174
Flanges, six flexible plastic, 97
Flavonols, 139
Flavor, 19,212
  aspiration, 109
  astringency vs. bitterness, 109
  buttery, 12
  definition of, 108-109
  finish, 109
Flavorant, 213
Food and wine combinations, 215,216
Fortified wines, 127
Fox grape (*Vitis labrusca*), 17
Fragrance (*nose*), 19-20,23,37,38,
  57,103-104
  balance, 103

Fragrance (*nose*) *(continued)*
  complexity, 103
  described, 103
  duration, 103-104
  harmony, 103
  identifying wine, 103
  qualities of, 103
  wine faults, 103
Fraud, detection, 128,129
French wines, 220
  Alsace, 87
Frost-damage, vineyards, 28
Fruit
  enzyme degradation, 15
  grape's taste, 14,32
Fungicides, modern, 57

Gall formation, 204
Gamay noir, grape variety, 37
  aroma, 102t
Garganega, aroma, 101t
Garrafeira, 21-22,85,87
Gastrin, 137
Geneva Double Curtain, 196
Geneva Research Station, 27,185
Geography
  light reflection off water, 32-33
  slope and aspect, 28-29,32
  soil drainage, 29
  soil type, 29-31
Geographic origin, 86-88
German wine(s), 23-24,86-87,176
  classification system, 24,177-178
  Mosel, 86
  nonchaptalized, 66
  Rheingau, 86
  sweetness, 178
Gewürztraminer
  from Alsace, 87
  aroma, 101t
  at Heron Hill, 71
  litchi nut aroma, 103

Glasses
  colorless, 95
  holding, 96
  ISO wine-tasting, 76,94,95i
Glenora Wine Cellars on Seneca
    Lake, 98
Glycerol, 65,181
  role in wine tears, 108
Gold Seal Winery, 147-163
Gout, 140-141
Governo process, 183
Grafting, 57,187
  benefits of, 189,202
Gran Coronas Black Label, 114
Grape(s)
  artificial taste, 32
  botrytis's source, 2
  cells, breaking down, 48
  composition of, 14,34
  growing, medieval Europe,
    195-196
  harvesting, 33-36
  history of growing, 30
  hybrid varieties, 17
  in Indo-European languages, 15
  Malvasia nero, 87
  need to be warmed, 37
  Negro amaro, 87
  riesling, 58,60
  solids, definition of, 60
  stomp, 40
  stored, 13
  taste of, 14
  varieties of, 17-18,23,35,184,188
Grapevine(s), 14
  breeding, 202,203
  buds, 194
  domestication, 14,16-18
    North America, 17
  fruiting potential, 193
  history of growing, 30-31
  nutritional needs, 200
  pruning, 28,29,192,196,199
    Australia, 196
    Europe, 196

Grapevine(s) *(continued)*
  self-fertilization, 192
  self-regulation, 198
  structure of, 192,193i,195i
  training, 30-31,194-197,197i
  in wild, 16-17
Greek wines, 172
Gringnolino, aroma, 102t
Ground covers, 201
Grünlack, 24

Hangover, 134
Harvesting, 33-36
  method, 35-36
  timing criteria, 33-34
Haziness, 99
Headache
  causes, 135-137
  prevention, 136
  types, 134,135-136
Health effects, of wine, 133-145
Health impact
  antimicrobial effects, 140
  antioxidant effects, 138-140
  bacteria, 140
  cancer, 141-142
  cardiovascular diseases, 137-138
  common cold, 140
  fetal alcohol syndrome (FAS),
    143-144
  gout, 140-141
  improved appetite, 137,143
  improved digestion, 137
  improved self-esteem, 137
  improved sleep, 137
    virus, 140
Hebarium plant press, 92
Heitz Nappa Valley Cabernet, 20-21
Heitz vineyards, 20
Herbicides, 199,201
Heron Hill Winery, 57
High density planting, 198
Histamine in wine, impact
    on headache, 134

Hotel school, 5
House wine, 208
Hydrogen sulfide, 113
Hydrometer, 47-48

Icewine (*Eiswein*), 116
Indo-European language, 15
Insecticidal soaps, 202
Iran, evidence of early wine making,
    16
Irrigation, 204-205
ISO (International Standards
    Organization) glasses,
    148,222
Isotope, 129-130
Isotopic ratios, 129,130
Italian wines, 9,23,87,88,120,157
  recioto, 182
Ithaca, 1,49,207
Ithaca burgers, 72

Joseph Phelps Late-Harvest
    Riesling, 131
Juice
  accidentally fermented, 13
  contaminated with iron, 45
  free-run, 48,49,50,62
  from stemmer-crusher at Herndon
    Hill, 58
  grape, 2,42,45

Kabinett, 178,179
Knot on cord technique, 97

Label(s), 21
  eagle symbol, 86
  German wine, 177
  removal, 92
  reserve, 85
  special terms, 21-22
  terms on, 85
  varietal origin, 85

Labrusca shoots, 196
Late-bottled vintage ports, 150
Leaf-galling, 189,190i
Lees, 52,176
Lenoir grape, 17
Lenticle(s), 167-168
Libby®#8470 glasses, 94
Lindemans Pyrus, 145
Lipoproteins, 138
Lopez de Heredia, 221
Lyre system, 196,197i

Magazines, 18
Malic acid, 15
    fermentation, 68-69
Malvasia nera, grape variety, 23
Manure, 199,200
Maréchal Foch, 37,38,39i,49,188
Marqués de Cáceres, 76
Marqués de Murrieta, 76
Maturation
    in oak, 70
    in stainless steel, 70
Mayacamus Chardonnay, 220
Medication and wine, 142
Mercaptans, 113
Merlot, aroma, 102t
Metabisulfite, 52,112
Metallic sensation, 113
Methyl anthranilate, 31-32
Microbial spoilage, 155
Migraine, 134
Minimal pruning, 196
Molinara, grape variety, 184
Monoamine oxidase inhibitors, 142
Monoterpenes, 123,124
Montrachet Chardonnay, 220
Montrose Estate Mudgee
        Chardonnay, 19
Mumm's Brut, 208
Muscat
    grape variety, 35,157
    wine aroma, 101t
Must, 62

Nature, 156
Nature-identical flavors, 129
Nebbiolo, 90
    aroma, 102t
Nederburg Edelkeur, 91
Negro amaro, grape variety, 23
Nematodes (roundworms), 202
Nerello mascalea, aroma, 102t
Niagara grapes chaptalization, 66
Noble, grape variety, 17
Noble rot, 24,178,181
Noradrenaline, 135
Norton, grape variety, 17
Nose (fragrance), 19

Oak
    barrel, 38,173
    casks, Heron Hill, 68
    flavor of, 70,173
    as seasoning, 70
    structures and properties, 172-173
Oak tree(s)
    gummy sap exudate, 14
    variation in species, 70,173
    variation of drying, 70
    variation of wood growth, 70,173
    white, 173
Off-odor, 41,111,112,118,125,
        170,204
Oloroso production, 152,154
Oregon wine industry, 22
Organic viticulture, 199,204
Ovary of grape vine, 192
Oxidation, 65
    off-odor, 41
Oxidized wine, 111,112

Parellada, aroma, 101t
Pasteurization, 125
Pest(s)
    control, biological, 202
    insects, 202
Pesticides, 199,202,203

Phenolic. *See also* Tannin
   antioxidant action, 41
   headache, 135
Phenolic compounds, 135
Phenolic fragrance, 184
Phenols, wine, 138
Phenolsulfotransferase (PST), 135
Phylloxera (root louse), 57,188,
      189,202
   outbreak in California, 191
Pigments, 67
   from stemmer-crusher, 5,58
Pinot blanc, wine aroma table, 101t
Pinot Grigio
   aroma, 101t
   from north Italy, 87
Pinot noir
   aroma, 102t
   from Burgundy, 87
   grape variety, 18,22
   from Oregon, 87,90
Placebo, 135
Planting, high density, 198
Plate filter, 68
Platelet sensitivity, 135
Pneumatic press
   disadvantage, 62
   operation of, 61-62,63i
Pollen, 16-17
Pomace, 50
Porcini mushrooms, 104
Port(s), 88,148-150
   color of, 149
   North American, 149
   Portuguese, 149-150
   production, 149
   types of, 150
   vintage, 104
Portuguese wines, 21-22,121
Prädikat, 24,178,179
   nonchaptalized, 66
Press cake, 50,61
Press head, 43
Presses, 37,42,43,44i
Prostaglandins, 136

Pruning systems, 196,197i
Punt, 96

QmP, nonchaptalized wines, 66
Qualitätswein, 179
Quality
   artistic, 116,121
   historic, 116,118
   subjective, 116-117
   of wine, determining, 116
Quercetin, 139,140,141

Racking, 69,175
   sediment remaining after, 99-100
Radioactive carbon, 129
Rancio, 149
Recioto della Valpolicella Amarone,
      183
Recioto Valpolicella, 182
Recioto wines, 182-183
Red wine(s), 126i,127
   aged, 87
   aroma, 102t
   fruity, 87
   headache, 136
   pigment, 15,35,45-46
   temperature for serving, 107-108
Redhead, 135-136
Refractometer, 34
Regional seasoning, 213
Reserve, on labels, 85
Restina, 172
Resveratrol, 138
Rheingau, region of Germany, 181
Riddling, 158,159
Riesling, 58,60,90
   aroma, 101t
   fragrance of, 57
   grape variety, 17,24,39i,40,57
   at Heron Hill, 71
Riesling, Joseph Phelps
      Late-Harvest, 131
Rioja, 9,76,87,221

Rondinella, grape variety, 184
Root galls, 189
Root louse (*Phylloxera*), 188
Rootstock, 187,190
    selection, 190-191
Rotten onion odor, 113
Rousanne, aroma, 101t
RT2T, 196,197i
Ruby port, 150
Rupestris rootstock, 189

Sanderson, Kip, 10
Sangiovese, aroma, 102t
Sauerkraut odor, 113
Sauvignon blanc, 87,90
    aroma, 101t
Schloss Johannisberg's Riesling, 24
Scion, 187,188
Scott Henry Trellis, 196,197i
Screwpull®, 97
Scuppernong, grape variety, 17
Sec (dry), 156
Second fermentation
    commercial, 69
    home made, 69
Sediment, taste of, 100
Semillon, aroma, 101t
Sensory
    analysis, professional, 103
    evaluation, 93
Serotonin, 135
Seyval blanc, 188
Sherry(ies), 88,150-155
    grape varieties, 152
    history of, 150
    New York, 151
    production of, 151-152,153i,154
    Spain, 151,152
    types, 152,154
    walnut aroma, 103
Shiraz, from Australia, 87
    aroma, 102t

Shoot(s), 187
    of grapevines, 192,196
    growth, 196-197
Skin contact, 98,99,127,140
Slope, 28-29
Soil
    drainage, 29
    erosion, 29
    friability, 201
    nutrition, 31
    structure, 30-31,201
    warming, 28-29
Solera, 151,152,154,155
Sotolon, 181
Sourness, 111
    detected in mouth, 106
    duration, 106
    produced by grapes, 67
    of wine, 106
South African wines, 91
Spanish wines, 87,114,221
    Marqués de Cáceres, 76,221
    Marqués de Murrieta, 76
    Rioja, 76
Sparkling wine, 155-156. *See also*
        Champagne
    and carbon dioxide, 136
    dosage, 158
    grapes used in, 157-158
    history of, 156
    production method
        charmat, 159-160
        traditional, 158
        transfer, 158,159
    sweetness, 155,156
    tirage, 158
Spätlese, 24,178
Spurs, 192
Statler Inn, 165,166
Stemmer-crusher, 45,58
    commercial operation
        of, 58,59i,60
Sticking, 60
Stoppers, plastic, 161-162

Sugar
added, 66
content, 127
fermentation, 50
in grapes, 14,34
limits on adding, 67
Sulfur, 111
Sulfur dioxide, 51,52,60-61,112,127
and asthmatics, 61
antimicrobial, 61
antioxidant, 61
medical problems, 61
medical risk, 141
synthesis by yeast, 61
Sur lies, 52
Suspension of sediment, 99
Sustained interest
definition of, 104
quality of fragrance, 103-104
Sweat socks odor, 113
Sweet wines, temperature
for serving, 107
Sweetness
detected in mouth, 106
duration, 105-106
produced by grapes, 67
of wine, 105-106
Syrah, 90
aroma, 102t
from Australia (Shiraz), 87
from California, 87

Tannin(s), 50,67,123,127
antioxidant action, 41
bitterness and astringency, 109
headache cause, 135
polymers, 135
from stemmer-crusher, 58
wine taste, 106-107
Tataric acid, 15
Tartrate crystals, 99
Tawny port, 150
TDN (1,1,6-trimethyl-1,2-dihydron-
aphthalene), 124

Temperature(s)
effect on wine, fermentation,
40,43,51,64,65
for serving wine, 107-108
Tempranillo, 90,221
aroma, 102t
Terrior, 30
Tio Pepe sherry, 218
Tirage, 158
Toasting of barrels, 173-174
Topping, 175
Torbata, aroma, 101t
Touriga nacional, aroma, 102t
Training of vine, 30,31,194-199
Tricyclic antidepressants, 142
Trockenbeerenauslese (TBA), 178,182
2-,4,6-TCA, 112. See also Corky;
Off-odor
Tyramine, headache, 134

United States, wine grown in, 20,22

Valpolicellas, 120
Varietal origin, 85-86,129
VDP (Vereinigung Deutscher
Prädikats-u
Qualitätsweinguter), 86
Venture Vineyard, 27-36
Vidal blanc, 188
Vine
age, 191
bisexuality, 16-17
growth, 195-196
training, 194-195
Vinegar bacteria, 41,64
Vinegary (off-odor and off-taste),
64,111,112
Vineyard(s)
establishing location of, 129
Geneva Research, 185,197i
Lopez de Heredia, 221
Marqués de Cáceres, 76
Marqués de Murrieta, 76
new world planting, 196

Vineyard(s) *(continued)*
 replanting cycles, 191
Vinho Tinto, 121
Vinho Tinto Garrafeira, 21-22
Vinifera shoots, 196
Vinifera Wine Cellars, 57
Vinifera Wine Cellars Riesling
  Icewine, 222
Vino santo, 183
Vintage
 avoiding bad years, 82
 avoiding regions with varied
  climates, 82
 charts, 82
 date, 88
 ports, 150
 techniques establishing, 199
Viticulture, organic, 199
Vitis labrusca, 17
Vitis riparia, 190
Vitis rupestris, 190
Vitis vinifera, 14,17,57,155,188
Viura, 90
 aroma, 101t
Volatile, definition of, 123

Weisslack, wine category, 24
White oaks
 American, 173
 French, 173
White wine(s)
 acidic, 127
 aroma, 101t
 commercial fermentation
  temperature, 65
 dry, 87
 from Germany, 86-87
 making at home, 40-45
 temperature for serving, 107-108
Willm Gewürztraminer, 221-222
Willmes pneumatic press, 60,61
Wine, 2
 adulteration, frequency, 128,130
 appearance, 98-100

Wine *(continued)*
 aroma, 100,103,101t-102t
 atypical flavors, 128
 blended, 80
 bottles
  half, 89
  screw-cap, 89
 breathing, 120
 Burgundy, 81
 buying, 80-81
 cellar rack, home, 77,78i
 and cheese, 216
 and children, 144-145
 contaminants, 170
 criteria for selection, 81
 critics, 119
 diamonds, 99
 drug interactions, 142
 establishing composition of, 129
 experts, 118-119
 faults, 99,110-114
  fragrance, 103
  misinterpretations, 111
  recognition of, 110
  various types of, 111
 finish, 19,20
 flavors of, 215
 and food, 20
 food value, 142-143
 fragrance, 100,103-104
 fraud, 128-131
 highly advertised, 80
 in Indo-European languages, 15
 magazines, 18,29
 and medications, 142
 old and breathing, 120
 old and fragrance, 120
 origin, 13-16
 presses, 37,42,43,44i
 quality of, 103
 regional suggestions, 86-88
 relationship with food, 214
 selecting in past, 209-210
 siphoning of, 53i
 swirling in the glass, 110

Wine *(continued)*
    smelling of, 104
    and taxation, 209
    toxic substances, 128
    upper-class, 81
    used in cooking, 210-212
    variable pricing, 81-82
    with dinner, 110
Wine Law of 1971, Germany, 177
Wine making, 27-31,36-47
    carbonic maceration, 36-40,48
    commercial, maturation, 69
    home, 14,36-47,39i
    origin, 13-16
    traditional
        red (red table), 39i,45-47,49
        white (white table), 39i,40-45
Wine tasting, 18-26,105-108
    bitterness, 106
    finish, 109
    flavor, 108-110
    glasses, 19
    procedure, 105
    saltiness, 106
    sensations
        mouth-feel, 19-23,37
        odor, 19-23,37
        taste, 15,19-23,34,38
        visual, 19-23
    sequence and intensity, 105
    sourness, 106
    sweetness, 105-106
    tannins, 106
    temperature for serving, 106-107
    toothpaste, 106
    water, 106

Wine tasting *(continued)*
    with food, 106
Wine tears, 108,172
Wine types from
    Australia, 145
    California, 22
    France, 87,88,220,221-222
    Germany, 176
    Greece, 172
    Italy, 87,88
    Portugal, 121
    Spain, 87,114,221

Yeast
    and acidity, 15
    additive, 45
    and aging potential, 67
    bloom on grapes, 33
    contribution to wine production,
        67
    distribution world wide, 14
    natural habitat, 14
    production of, 40,41
        fragrant compounds, 67
        taste substances, 67
    role in wine production, 67
    selection during fermentation,
        14,67
    spoilage, 41,60
    types of, 14
    wild, 68

Zinfandel, aroma, 102t